HUMAN-COMPUTER INTERACTION, 2002, Volume 17,
Copyright © 2002, Lawrence Erlbaum Associates, Inc.

Introduction to This Special Issue on Text Entry for Mobile Computing

I. Scott MacKenzie
York University

The four articles in this special issue of *Human–Computer Interaction* describe recent research in mobile text entry. Text entry research is by no means new. We can depict the progression of text entry research as occurring in two waves. The first was in the 1970s and early 1980s in response to the new role of electronic computers in automating office tasks, such as typing, word processing, and document management. Modeling 10-finger touch typing, categorizing typing errors, task analysis, and comparing document editing strategies are some of the themes in this early research. Exemplary references are Cooper (1983) and Chapters 3 to 9 in *The Psychology of Human–Computer Interaction* (Card, Moran, & Newell, 1983).

The second wave of text entry research was not aimed at two-handed input in the workstation setting but at the more demanding environment of mobile computing. This began with the arrival of pen-based systems in the early 1990s. The automated recognition of handwritten text entered with a stylus was the elixir for this new mode of interaction, it seemed. Although the potential of pen-based computing, and mobility in general, was well beyond the recognition of hand-printed symbols, this aspect of the interaction received much of the attention. And the attention was not good. Promises did not meet the expectations of demanding users, and the market suffered significantly for this. Yet products continued to arrive. The single most significant event in pen-based computing was the introduction in 1995, and tremendous success of, the Palm™ Pilot from Palm, Inc. (now a division of 3Com). For text entry, Palm sidestepped many of the problems with existing handwriting recognition technology. To avoid the segmentation problems inherent in multistroke hand-printed or cursive text, Palm introduced Graffiti®, a commercial realization of the one-stroke-per-symbol technique known as Unistrokes (Goldberg & Richardson, 1993). For users wanting to avoid handwritten text entry altogether, the Palm Pilot and other pen-entry devices also included a soft keyboard allowing users to tap on virtual keys, thus directly obtaining the desired symbol.

Research in stylus-based text entry is ongoing; however, this second wave now includes significant interest in text entry using small physical keyboards. Some devices allow text entry with as few as five keys, such as the AccessLink® II by Glenayre Electronics (Charlotte, NC). Others sport a complete, but miniature, Qwerty keyboard, such as the BlackBerry™ by Research in Motion (Waterloo, Canada). These are both examples of two-way pagers. As well, text entry using the mobile phone keypad has recently grabbed the attention of users and researchers. The success of so-called text messaging on mobile phones is nothing short of remarkable. The ability to send a message from one phone to another discretely, asynchronously, and at very low cost has proven hugely successful, particularly in Europe. The statistics are remarkable: more than a billion text messages per month! This is particularly surprising in view of the limited capability of the mobile phone keypad for text input. Not surprisingly then, the current wave of mobile text entry research includes numerous researchers and companies working on new ideas to improve the text entry technique for mobile phones or other anticipated mobile products supporting similar services.

This special issue opens with a review article by MacKenzie and Soukoreff. The article begins with general comments on methodologies for evaluating new text entry techniques, including an elaboration of the differing attention demands between text-creation tasks and text-copying tasks. Other aspects of evaluation include the need to consider both the novice experience and the expert potential of a proposed technique. Some of the more recent and exciting research in mobile text entry involves the combined use of movement minimization techniques and language modeling in designing and optimizing the text entry task. Many factors in this area are noted and elaborated, with particular attention to the ambiguity in the telephone keypad, where each key encodes either three or four letters. The article then surveys text entry techniques for mobile computing, including numerous key-based and stylus-based techniques.

Emerging from the language theory seeded in Shannon's early work (Shannon, 1951; Shannon & Weaver, 1949), Ward, Blackwell, and MacKay devised a dynamic interface for text entry. In the second article in this special issue, they present Dasher, a text-entry technique with an interface driven from continuous two-dimensional gestures that regulate the flow of textual information across the display. The textual information is organized to fully exploit language redundancy, thus affording the means to accelerate the text entry task. For example, having just entered t, a set of follow-on letters progresses across the screen with more display space given to more likely letters, thus facilitating selection of, for example, h, and, subsequently, e, in entering the. Of course, the details are everything, and in this Ward et al. cover their territory well. They describe mechanisms for correcting errors and for driving the interface with devices such as a stylus or eye-tracker. Refreshing, as well, is their detailed extension of their interface using the Hiragana character set used in Japanese text entry.

In the third article, Zhai, Hunter, and Smith provide a detailed extension to the prediction model of Soukoreff and MacKenzie (Soukoreff & MacKenzie, 1995). The original motivation was MacKenzie and Zhang's description and evaluation of a high-performance soft keyboard designed using this model (MacKenzie & Zhang, 1999). MacKenzie and Zhang designed their keyboard by trial and error. Specifically, the model was embedded in a spreadsheet that included digram probabilities from a language corpus and coefficients for a movement time prediction model. A separate collection of cells contained characters representing a soft keyboard. Formulae in the spreadsheet collected together the various components of the model and produced an estimate of the expert text entry rate for the given soft keyboard. By directly editing the soft keyboard cells, different layouts were tested with a prediction appearing immediately with each edit. MacKenzie and Zhang worked the spreadsheet using simple heuristics to fine-tune the layout (e.g., common letters should be clustered together near the center; push infrequent letters to the edges). Predictions edged upward, and eventually a design, Opti, was settled on and tested in a user study. Zhai et al. immediately spotted an opportunity in this methodology. Instead of manually rearranging letters, why not use an algorithm to automatically explore the design space? In their article, they present two quantitative techniques to search for optimized virtual keyboard layouts. The first technique simulates the dynamics of a keyboard with "digraph springs," producing a Hooke keyboard. The second yields a Metropolis keyboard using a random walk algorithm guided by a "Fitts-digraph energy" objective function. The details of these design techniques represent a welcome addition to research in mobile text entry.

In the final article, Hughes, Warren, and Buyukkokten take a completely different approach to modeling the text entry task. By conventional practice, a model is built under a set of conditions that are limited in number, yet are representative of the broader range of conditions under which the model is later applied. Fitts' law is an example of this approach (Fitts, 1954). Typically, the model is built using a representative set of conditions, say, eight different combinations of target width and amplitude, and then is later applied under conditions never actually used in building the model. Such is the generality of Fitts' law that the subsequent predictions are surprisingly accurate. Hughes et al. instead built an empirical bi-action table containing an entry for each and every condition. Each entry is an empirical measurement of the time for users to make an action, given a preceding action. Despite the apparent brute-force approach, there is indeed a modeling component to their efforts. During the data collection stage, the actions bear no linguistic assignment. The actions are simple motor acts, performed repeatedly at peak rates for short durations. Armed with a table of minimum movement times for the action set, they proceeded with the subsequent task of assigning letters to keys and searching for the optimal assignment. Thus, they too arrive at an optimal design but the journey is along a very different path.

As a collection, these articles represent a sample of promising research initiatives in text entry for mobile computing. We hope you find the work both stimulating and suggestive of further avenues to explore in this exciting research domain.

NOTES

Acknowledgment. We thank the reviewers who participated in the process of bringing this collection of articles to print, as well as Thomas Moran, Editor of *Human–Computer Interaction,* for suggesting the theme of this special issue.

REFERENCES

Card, S. K., Moran, T. P., & Newell, A. (1983). *The psychology of human–computer interaction.* Hillsdale, NJ: Lawrence Erlbaum Associates, Inc.

Cooper, W. E. (1983). *Cognitive aspects of skilled typewriting.* New York: Springer-Verlag.

Fitts, P. M. (1954). The information capacity of the human motor system in controlling the amplitude of movement. *Journal of Experimental Psychology, 47,* 381–391.

Goldberg, D., & Richardson, C. (1993). Touch-typing with a stylus. *Proceedings of the INTERCHI 93 Conference on Human Factors in Computing Systems.* New York: ACM.

MacKenzie, I. S., & Zhang, S. X. (1999). The design and evaluation of a high-performance soft keyboard. *Proceedings of the CHI 99 Conference on Human Factors in Computing Systems.* New York: ACM.

Shannon, C. E. (1951). Prediction and entropy of printed English. *Bell System Technical Journal, 30,* 51–64.

Shannon, C. E., & Weaver, W. (1949). *The mathematical theory of communications.* Urbana: University of Illinois Press.

Soukoreff, W., & MacKenzie, I. S. (1995). Theoretical upper and lower bounds on typing speeds using a stylus and soft keyboard. *Behaviour & Information Technology, 14,* 370–379.

ARTICLES IN THIS SPECIAL ISSUE

Hughes, D., Warren, J., & Buyukkokten, O. (2002). Empirical bi-action tables: A tool for the evaluation and optimization of text input systems. Application I: Stylus keyboards. *Human–Computer Interaction, 17,* 271–309.

MacKenzie, I. S., & Soukoreff, R. W. (2002). Text entry for mobile computing: Models and methods, theory and practice. *Human–Computer Interaction, 17,* 147–198.

Ward, D. J., Blackwell, A. F., & MacKay, D. J. C. (2002). Dasher: A gesture-driven data entry interface for mobile computing. *Human–Computer Interaction, 17,* 199–228.

Zhai, S., Hunter, M., & Smith, B. A. (2002). Performance optimization of virtual keyboards. *Human–Computer Interaction, 17,* 229–269.

HUMAN-COMPUTER INTERACTION, 2002, Volume 17, pp. 147–198

Text Entry for Mobile Computing: Models and Methods, Theory and Practice

I. Scott MacKenzie and R. William Soukoreff
York University

ABSTRACT

Text input for mobile or handheld devices is a flourishing research area. This article begins with a brief history of the emergence and impact of mobile computers and mobile communications devices. Key factors in conducting sound evaluations of new technologies for mobile text entry are presented, including methodology and experiment design. Important factors to consider are identified and elaborated, such as focus of attention, text creation versus text copy tasks, novice versus expert performance, quantitative versus qualitative measures, and the speed–accuracy trade-off. An exciting area within mobile text entry is the combined use of Fitts' law and a language corpus to model, and subsequently optimize, a text entry technique. The model is described, along with examples for a variety of soft keyboards as well as the telephone keypad. A survey of mobile text entry techniques, both in research papers and in commercial products, is presented.

I. Scott MacKenzie is a computer scientist with an interest in human–computer interaction, especially human input to computing systems and human performance measurement and modeling; he is an Associate Professor in the Department of Computer Science at York University. **R. William Soukoreff** is a computer scientist with an interest in human–computer interaction; he is a graduate student in the Department of Computer Science at York University.

CONTENTS

1. INTRODUCTION

Although text entry is by no means new in mobile computing, there has been a burst of research on the topic in recent years. There are several reasons for this heightened interest: First, mobile computing is on the rise and has spawned new application domains such as wearable computing, two-way paging, and mobile Web and e-mail access. Second, word processors, spreadsheets, personal schedulers, and other traditional desktop applications are increasingly available on mobile platforms. Third, there is a strong demand for the input of text or alphanumeric information that is easily and efficiently entered, recognized, stored, forwarded, or searched, via traditional software techniques. Fourth, the phenomenal success of text messaging with mobile phone users has inspired considerable speculation on future spin-off technologies, all expected to benefit from text entry.

The statistics for text messaging on mobile phones are remarkable. In January 2001, GSM Europe reported that 15 billion Short Message Service (SMS) text messages are transmitted per month worldwide.[1] This is particularly interesting in view of the limited capability for text input with the current generation of mobile phone technology.

Although the ubiquitous Qwerty keyboard reigns supreme as the primary text entry device on desktop systems, mobile and handheld systems lack an equivalent dominant technology or technique for the same task. And so, the challenge of text entry for mobile computing presents itself. A valid question is, Why not just apply the Qwerty keyboard to the mobile paradigm? Despite the obvious advantage of familiarity, a Qwerty keyboard is bulky, and unless the keyboard is full size, touch typing is hampered or impossible. In addition, some mobile devices are intended for one hand use, and this reduces the advantage of the Qwerty arrangement. (A notable exception is the *Half-Qwerty* keyboard, discussed later.) Many mobile devices are committed to the pen input paradigm, so a Qwerty keyboard is simply not an option. Where physical buttons or keys are employed, the mobile form factor often limits the key complement to a dozen or fewer keys.

This article is organized as follows. We begin with a brief historical background of mobile and handheld devices. This is important because it juxtaposes the efforts of researchers with the corporations that created early mobile and handheld devices. Following this, we offer some opinions and observations on the evaluation of text input techniques. Many, but not all, of the techniques described later in this article have been empirically evaluated in user tests. To compare input technologies, the results of these evaluations are crucial. Factors to consider are presented and elaborated. Following this, we detail one of the most active areas of current research—optimization of text entry using language and motor control modeling. Finally, we present a survey of the current state of the art in text entry for mobile computing. We conclude with some observations on the technologies reviewed and the open research questions that remain.

---

1. GSM stands for Global System for Mobile communications. The GSM Association, based in Dublin, Ireland, represents the interests of hundreds of satellite operators, manufacturers, suppliers, and regulatory and administrative bodies from around the world. See http://www.gsmworld.com for further details.

There are two notable omissions in this article. One is speech recognition as a vehicle for text entry. Always "about to emerge," speech is an input technology quick to grab headlines but perennially unable to enter the mainstream of computing. In our view, speech is a deserving (albeit niche) technology, but it is unlikely to supplant traditional interaction techniques for desktop or mobile computing (see Shneiderman, 2000, for further discussion). Although some mobile phones support limited speech recognition, the interaction consists of selecting from a small list of preprogrammed entries, such as names in an address book. Speech recognition is not used for general purpose text input on mobile devices.

The other omission is international languages. It is clear and obvious that text entry does not imply "English text entry." Languages throughout the world are currently supported in various forms in mobile computing, and this will continue. Although the focus in this article is on English, the discussions apply to other languages, particularly those based on the Roman alphabet (see Sacher, 1998, for a discussion on text entry in Asian languages).

1.1. Mobile Computing

Among the earliest of handheld devices was the HP95LX, which was released in 1991 by Hewlett-Packard (Palo Alto, CA; http://www.hp.com). The technological equivalent of an IBM® XT shrunk into a clamshell format, the HP95LX was small enough to fit in the palm of one's hand. Although the term *Personal Digital Assistant* (PDA) had not yet entered the vernacular to describe a handheld computer, this was the first PDA. The HP95LX provided a small Qwerty keyboard for text entry, although touch typing was impossible due to its size. Later devices (the HP100LX and HP200LX) followed. These devices demonstrated that the Qwerty keyboard could be adapted to mobile computing devices.

The early 1990s was an exciting time for mobile computing due to the arrival of pen computing. The ideas touted much earlier by Kay and Goldberg (1977) in their Dynabook project finally surfaced in commercial products. However, the initial devices were bulky, expensive, and power-hungry, and they could not deliver in the one area that garnered the most attention—handwriting recognition. Without a keyboard, the pen was the primary input device. If only "selecting" and "annotating" were required, then the success of pen entry seemed assured. However, some applications demanded entry of text as machine-readable characters, and the handwriting recognition technology of the time was not up to the challenge. Products from this era, such as GRidPaD, Momenta™, Poqet®, and PenPad, did not sustain the volume of sales necessary for commercial viability. Most endured only 1 or 2 years.

One of the most significant events in pen-based computing was the 1993 announcement from Apple® Computer, Inc. (Santa Clara, CA; http://www.apple.com) that it would enter the pen computing market. Thus emerged the Apple MessagePad® (a.k.a. Newton®). Apple was a major player in desktop computing at the time, and its commitment to pen computing was taken seriously. To a certain extent, Apple added legitimacy to this entire segment of the computing market. However, the Newton was expensive and rather specialized. It was embraced by many technophiles, but it did not significantly penetrate the larger desktop or consumer market. The Newton's handwriting recognition, particularly on early models, was so poor that it was ridiculed in the media—by Garry Trudeau (1996), for example, in his celebrated *Doonesbury* cartoons. Nevertheless, the Newton received considerable attention, and it ultimately set the stage for future mobile devices.

The next significant event in mobile or pen computing was the release in 1996 of the Palm™ Pilot (now called the Palm) by Palm Inc. (Santa Clara, CA; http://www.palm.com). The Palm was an instant hit. Five years hence, it is the technology of choice for millions of users of mobile devices. There is much speculation on why the Palm was so successful. Some factors seem relevant: The price was about $500, a few hundred less than a Newton. The Palm supported HotSync® (including cables and software for transferring data between the Palm and a desktop computer) as a standard feature. The Palm was smaller and lighter than the Newton, and could fit in one's pocket. Because of lower power consumption, the batteries lasted for weeks instead of hours. Finally, and perhaps most important, the Palm avoided the thorny issue of cursive or block-letter handwriting recognition by introducing a greatly simplified handwriting technique known as Graffiti® (which is discussed later in this article). By simplifying recognition, Graffiti required less CPU power and memory, achieved better character recognition, and ultimately enjoyed widespread acceptance among users.

The year 1996 also saw the release of the Windows® CE operating system by Microsoft® (Redmond, WA; http://www.microsoft.com). Devices such a Casio's Cassiopeia® or Philips' Velo, which used Windows CE, were more powerful than previous mobile computing devices, but were also larger. The first version of Windows CE only supported a soft keyboard for text entry, but later versions included the JOT handwriting recognizer, by Communications Intelligence Corporation® (Redwood Shores, CA; http://www.cic.com), and Microsoft Transcriber.

A recent entry in the pen computing market is the CrossPad by A. T. Cross Company (Lincoln, RI; http://www.cross.com). The CrossPad avoids handwriting recognition by recording the user's writing as ink trails. The user's notes are downloaded to a desktop computer for storage and subsequent recognition on the desktop computer. Software accompanying the CrossPad supports handwriting recognition of keywords, and indexing and retrieval by keyword.

All of the devices previously mentioned are handheld computers that support text entry. Another quite different group of devices that support text entry are messaging devices such as mobile phones and pagers. In Europe, where text messaging has been available since 1991, more text messages are transmitted daily than voice messages.[2] In North America, most mobile phones and pagers do not yet support text messaging, but this is changing. The latest generation of two-way pagers such as the BlackBerry™ by Research In Motion (Waterloo, Ontario, Canada; http://rim.net, but also see http://www.blackberry.net) and PageWriter® by Motorola (Schaumburg, IL; http://www.motorola.com) support text entry via a miniature Qwerty keyboard.

We conclude this perspective with a hint at what the future might hold. At the top of our list is a device combining the programmability of the PDA, wireless telephony, text messaging, and unfettered Internet and e-mail access. Pieces of this scenario already exist, but implementations require a specialized configuration, optional components, or support only a subset of standard features. We view these as transitional technologies. Devices that do not quite make the grade, in our view, are those that require an add-on radio transceiver or provide Internet access only to sites supporting a specialized protocol (e.g., wireless access protocol).

For text input, the pen-based paradigm has dominated the PDA market, but there is a parallel trend toward text messaging in mobile phones and pagers using keyboard-based technology. If these technologies converge, then which text input technology will prevail? This is a difficult question to answer, and although there is no definitive answer, the following section identifies the key issues to consider.

1.2. Text Entry

There are two competing paradigms for mobile text input: pen-based input and keyboard-based input. Both emerged from ancient technologies ("ancient" in that they predate computers): typing and handwriting. User experience with typing and handwriting greatly influences expectations for text entry in mobile computing; however, the two tasks are fundamentally different.

---

2. SMS (Short Message Service) is the predominant text-messaging technology in Europe. SMS supports transmission and reception of messages up to 160 characters via mobile phones (phone-to-phone). *Instant Messaging* is a similar technology popular in North America, but it is used mostly in PC-to-phone messaging.

A key feature of keyboard-based text entry is that it directly produces machine-readable text (i.e., ASCII characters), a necessary feature for indexing, searching, and handling by contemporary character-based technology. Handwriting without character recognition produces "digital ink." This is fine for some applications such as annotation, visual art, and graphic design. However, digital ink requires more memory and in general it is not well managed by computing technology. Specifically, digital ink is difficult to index and search (although Poon, Weber, & Cass, 1995, reported some success with a graphical search mechanism for digital ink that is not based on recognition). For handwritten text entry to achieve wide appeal, it must be coupled with recognition technology.

An important consideration implicit in the discussion of text input technology is user satisfaction. The point was made earlier that the Palm succeeded where the Newton failed, in part because of users' acceptance of Graffiti as a text input technology. Users' expectations for text entry are set by current practice. Modest touch typing speeds in the range of 20 to 40 words per minute (wpm) are achievable for hunt-and-peck typists. Rates in the 40 to 60 wpm range are achievable for touch typists, and with practice, skilled touch typists can achieve rates greater than 60 wpm. Handwriting speeds are commonly in the 15 to 25 wpm range. These statistics are confirmed by several sources (Card, Moran, & Newell, 1983; Devoe, 1967; Lewis, 1999; MacKenzie, Nonnecke, Riddersma, McQueen, & Meltz, 1994; Van Cott & Kinkade, 1972). Users, perhaps unrealistically, expect to achieve text input rates within these ranges on mobile devices. Furthermore, they expect these rates immediately, or within a short time of using a new input technology.

The preceding paragraphs have outlined qualities of a successful text input method. Production of machine-readable characters at a speed acceptable to users is a reasonable objective. To determine if a particular text input method meets this objective, or to compare new and existing text input methods, a user evaluation is needed.

2. EVALUATION

Research in mobile text entry is flourishing in part because user needs are not currently met. Typically, traditional text input technologies are refined or new input technologies are invented. Either way, evaluation is a critical and demanding part of the research program. The questions researchers pose are ambitious: Can entry rates be improved if we arrange the buttons on a keyboard in a certain way? What is the effect if we use context to guess the next letter or word? Can we apply an altogether different technology, like pie menus, touch pads, or pattern recognition, to the problem of text input? In this section, we discuss important issues in undertaking valid and useful evaluations of text entry techniques.

2.1. Methodology

An evaluation is valuable and useful if the methodology is reproducible and results are generalizable. *Reproducible* implies that other researchers can duplicate the method to confirm or refute results. This is achieved for the most part simply by following an appropriate reporting style (e.g., American Psychological Association, 1995). *Generalizable* implies that results have implications beyond the narrow context of the controlled experiment. This is achieved through a well-designed experiment that gathers measures that are accurate and relevant, in tasks that are representative of real-life behavior. There is, unfortunately, a trade-off here. In real life, people rarely focus solely on a single task. Methodologies so designed, therefore, may find that the measurements include behaviors not specifically required of the interaction technique. The trade-off, therefore, is between the accuracy of our answers and the importance or relevance of the questions they seek to address. That is, we can choose between providing *accurate answers* to *narrow questions*, or providing *vague answers* to *broad questions*. The reader is implored not to interpret this too strictly, but, we hope, the point is made. In designing an experiment, we strive for the best of both worlds; answering interesting or broad questions (viz., using real-life tasks) and doing so accurately (viz., accurately measuring the behavior of interest, such as entry speed or accuracy).

In the following sections, we identify some factors relevant to methodologies for evaluating text entry on mobile systems.

2.2. Text Creation Versus Text Copy

An important distinction in text entry evaluations is between *text creation* and *text copy*. In a text copy task, the participant is given text to enter using the input technique under investigation. In a creation task, the source text is either memorized or generated by the participant. Although text creation is closer to typical usage, the approach is generally not appropriate for an empirical evaluation. This is explained in the following paragraphs.

As a backdrop for discussing these two types of tasks, we introduce the term *focus of attention* (FOA). FOA speaks to the attention demands of the task. Consider the case of an expert touch typist using a Qwerty keyboard to copy text from a nearby sheet of paper. This is a text copy task. Because the typist is an expert, she does look at the keyboard or display—she attends only to the source text. This is a single FOA task. However, if input is via a stylus and soft keyboard, the typist must also attend to the keyboard. (A soft keyboard cannot be operated "eyes free.") Stylus typing, therefore, is a two-FOA task. If the typist is at a less-than-expert level and corrects errors, she must look at the display to monitor results. This increases touch typing to a two-FOA task and stylus typing to a three-FOA task. Clearly, the feedback channel is overburdened in a three-FOA task.

Despite the additional FOA, text copy tasks are generally preferred to text creation tasks for empirical evaluations. There are several reasons. One is the possible presence of behaviors not required of the interaction technique. Examples include pondering ("What should I enter next?") or secondary tasks (e.g., fiddling with system features). Clearly, measurement of text entry speed is compromised if such behaviors are present.

A second difficulty with text creation tasks is identifying errors—it is difficult to know exactly what a participant intended to enter if the participant is generating the text. Even if the message content is known a priori, errors in spelling or memory recall may occur, and these meta-level mistakes are often indistinguishable from errors due to the interface itself.

A third difficulty is the loss of control over the distribution of letters and words entered. The task should require the participant to enter a representative number of occurrences of characters or words in the language (i.e., results are generalizable). However, it is not possible to control for this if the participant is generating the text.

The main advantage of a text creation task is that it mimics typical usage. The disadvantages just cited, however, are significant and drive most researchers to use text copy tasks despite the increased FOA. One way to mitigate the effects of increased FOA is to dictate the source text through the audio channel. Ward, Blackwell, and MacKay (2000) used this technique; however, they noted that participants found the approach stressful and hard to follow.

A carefully designed experiment may capture the strengths of both a text creation task and a text copy task. One technique is to present participants with short, easy-to-memorize phrases of text. Participants are directed to read and memorize each phrase before entering it. Entry proceeding thus benefits from the desirable property of a text creation task (viz., reduced FOA). In addition, the desirable properties of a text copy task are captured (i.e., control over letter and word frequencies and performance measurements that exclude thinking about what to write). There are numerous examples of this approach in the literature (e.g., Alsio & Goldstein, 2000; MacKenzie, Nonnecke, Riddersma, et al., 1994; MacKenzie & Zhang, 1999; Rau & Skiena, 1994). A similar technique is to present text in large a block (e.g., a complete paragraph) but to interleave each line of the presented text (input) with each line of generated text (output). As input proceeds, each character entered appears directly below the intended character. This is a text copy task; however, FOA is reduced to that of a text creation task because participants attend only to one location for both the source text and the results of entry. An example of this methodology is reported by Matias, MacKenzie, and Buxton (1993) and Matias, MacKenzie, and Buxton (1996a).

2.3. Novice Versus Expert Performance

Most work on the design of text input methods focuses on the potential, or expert, text entry rate of a particular design. However, the novice experience is paramount for the success of new text input methods. This is at least partially due to the target market. Mobile devices, such as mobile phones and PDAs, once specialized tools for professionals, are increasingly targeted for the consumer market. It follows that "immediate usability" is important. In other words, it may be a moot point to establish the expert, or "potential" text entry rate for an input technique if prolonged practice is required to achieve it. Consumers, discouraged by their initial experience and frustration, may never invest the required effort to become experts.

However, measuring immediate usability is easier said than done. In typical studies of new interaction techniques, participants are given a demonstration of the technique followed by a brief practice session. Then, data collection proceeds over several blocks of trials. However, the measurements are a poor indicator of novice behavior, at least in the sense of immediate, or walk-up, usability. Within a few minutes, participants' knowledge of the interaction technique develops, and the novice status fades. Measuring expert performance is also not easy because acquisition of expertise requires many blocks of trials administered over many days, or more.

Some longitudinal text entry studies are hereby cited (Bellman & MacKenzie, 1998; Gopher & Raij, 1988; MacKenzie & Zhang, 1999; Matias et al., 1996a; McMulkin, 1992). An example of results from a typical longitudinal study is given in Figure 1. Users' improvement in entry speed is shown over 20 sessions of input for two types of soft keyboard. The data were fitted to the standard power law of learning (see Card, English, & Burr, 1978). Prediction equations and squared correlations are shown, as are extrapolations of the predictions to 50 sessions.

2.4. Quantitative Versus Qualitative Analyses

We noted earlier a trade-off between the accuracy of answers and the relevance of the questions they seek to address. Quantitative evaluations tend to provide accurate answers to narrow questions, whereas qualitative evaluations tend to provide rather loose answers ("participants liked the device!") to broad but very important issues (comfort, ease of use, subjective impression, etc.). Of course, researchers strive for the best of both worlds. In quantitative evalua-

Figure 1. **Reporting example for a longitudinal study. From "The design and evaluation of a high-performance soft keyboard," by I.S. MacKenzie and S. X. Zhang, 1999,** *Proceedings of the CHI 99 Conference on Human Factors in Computing Systems.* **Copyright 1999 by ACM. Reprinted with permission.**

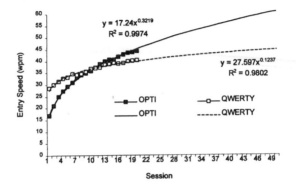

tions, "representative tasks" and "relevant measures" are used to ensure interesting or relevant questions are answered. In qualitative evaluations, robust test instruments are developed to ensure the answers are accurate, relevant, reproducible, and generalizable.

When reporting quantitative results, there are many common pitfalls to avoid such as inaccuracy in measurements, lack of control or baseline conditions, inferring too much from data, using too small a sample size, collecting insufficient data, artificially biasing data by aggregation, nonrandom presentation of conditions, and inappropriate treatment of outliers. The reader is directed to textbooks in experimental psychology for further discussions (e.g., Martin, 1996; for a discussion on aggregation bias, see Walker et al., 1993).

Researchers may be excused for slightly bending the rules, perhaps, but all too common are published reports stating only qualitative results steeped in anecdote, or, worse yet, testimonials unsupported by empirical data. An excerpt from one such publication illustrates our point:

> While we have yet not done systematic user testing, anecdotal experience to date is consistent: Users well practiced in both ... and ... consistently find the latter to be about three times faster, with accuracy for both systems very high.[3]

Testimonials such as this are of questionable merit; they surely do not meet the criteria for good research—that results are generalizable and reproducible.

---

3. An excerpt from a paper published in the proceedings of a conference in human–computer interaction.

Unless a controlled experiment is performed using quantitative metrics or established qualitative test instruments, there is no way to gauge the performance of a new text input technique. Conjuring up a new input technique is fine, but research demands more. It demands that new ideas are implemented and evaluated in conformance with the rigors of an empirical evaluation.

Although quantitative tests form the backbone of any scientific study, qualitative aspects of the investigation are also important. In human–computer interfaces, users must feel comfortable with the interaction technique and must feel their efforts have a reasonable payoff in their ability to accomplish tasks. Participants will develop impressions of each device or condition tested, and these should be solicited and accounted for in the final analysis. Typically, these opinions are sought via questionnaire, administered at the end of a condition or experiment. The reader is referred to textbooks in human–computer interaction for direction in questionnaire design (e.g., Dix, Finlay, Abowd, & Beale, 1998).

2.5. Speed

For text input there are two primary evaluation metrics: speed and accuracy. The simplest way to measure and report speed is to measure the number of characters entered per second during a trial, perhaps averaged over blocks of trials. This gives a measure in characters per second (cps). To convert this to wpm, the standard typists' definition of a word as five characters (regardless of whether the characters are letters, punctuation, or spaces) is employed (Gentner, Grudin, Larochelle, Norman, & Rumelhart, 1983). Therefore, wpm is obtained by multiplying characters per second by 60 (seconds per minute) and dividing by 5 (characters per word).

2.6. Accuracy

Accuracy is more problematic. For a simple treatment of accuracy, we obtain a metric that captures the number of characters in error during a trial and report these as a percentage of all characters in the presented text. A more complete analysis involves determining what kind of errors occurred, and why. The difficulty arises from the compounding nature of mistakes (see Suhm, Myers, & Waibel, 1999), and the desire to automate as much of the data measurement and analysis as possible. Four basic types of errors include entering an incorrect character (substitution), omitting a character (omission), adding an extra character (insertion), or swapping neighboring characters (transposition). Although it is straightforward for a human to compare the intended text with the generated text and tabulate the errors, in practice the amount of analysis is simply too much, given a reasonable number of participants, conditions, and trials. Additionally, tabulation errors may be introduced if performed manually.

However, automating error tabulation is not trivial. Consider an experiment where the participant is required to enter the 19-character phrase "the quick brown fox." If the participant enters "the quxxi brown fox," the incorrect word contains either three substitution errors or two insertion ("xx") and two omission ("ck") errors. The explanation with the fewest number of errors (3) is preferred and, in this simple example, yields an error rate of (3 / 19) × 100% = 15.8%. Algorithms for "string distance" calculations, such as the Levenshtein string distance statistic (Damerau, 1964; Levenshtein, 1966), might assist in automating analyses such as these, as demonstrated by Soukoreff and MacKenzie (2001).

Difficulties in error tabulation have pushed some researchers to ignore errors altogether (e.g., Venolia & Neiberg, 1994) or to force the participant to enter correct text only (e.g., Lewis, 1999).

Directing participants to "correct as you go" is another possible approach. Assuming participants adhere to instructions, the resulting text is error free; thus, the error rate is 0%. In general though, participants will leave errors in the generated text, even if requested not to. The result is two levels of errors—those that were corrected and those that were not. For the corrected errors, overhead is incurred in making the corrections. A reasonable measure of accuracy in this case is keystrokes per character (KSPC). For a Qwerty keyboard, the ideal is KSPC = 1.0, but, in practice, KSPC > 1 if participants correct as they go. If, for example, a 25-character phrase were entered and two substitution errors occurred, each corrected by pressing Backspace followed by the correct character, then KSPC = (25 + 4) / 25 = 1.16.[4]

A useful tool for designers is the *confusion matrix,* graphically depicting the frequency of character-level transcription errors. Figure 2 is a confusion matrix taken from MacKenzie and Chang's (1999) comparative study of two handwriting recognizers. The confusion matrix displays intended characters versus recognized characters illustrating how often an intended character (left-hand column) was misrecognized and interpreted as another character (bottom row). Each dot represents three occurrences.

Clearly, both speed and accuracy must be measured and analyzed. Speed and accuracy are commonly known to exist in a continuum, wherein speed is traded for accuracy and vice versa (Hancock & Newell, 1985; Pachella & Pew, 1968; Pew, 1969; Swensson, 1972; Wickelgren, 1977). Participants can enter

---

4. Correct-as-you-go has an additional problem: reaction time. If entry proceeds quickly, an error may be followed by several additional entries before the participant can react to the error. The overhead in correcting the error may be substantial (see Matias, Mackenzie, & Buxton, 1996a, for a discussion of this).

Figure 2. Sample confusion matrix. From "A performance comparison of two handwriting recognizers," by I. S. MacKenzie and L. Chang, 1999, *Interacting with Computers, 11*, pp. 283–297. Copyright 1999 by Elsevier Service. Reprinted with permission.

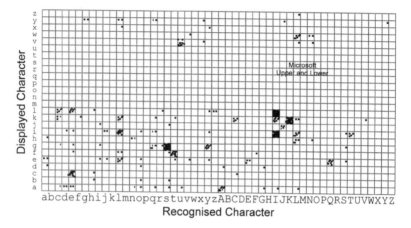

text more quickly if they are willing to sacrifice accuracy. For participants to perform with high accuracy, they must slow down. The trade-off suggests that measuring only speed or only accuracy will skew the results so as to make the text input method appear better (or worse!) than it really is. An example of a reporting technique that combines speed and accuracy is given in Figure 3. Conditions are "better" toward the top and right of the figure because they are both fast and accurate.

2.7. Other Factors

The text input process can be significantly impacted by factors bearing little on the input device, such as whether the device is operated standing, sitting, or walking, or whether it is operated with one or two hands. Designers of novel text input techniques must be aware that users want to operate mobile devices anytime, anywhere. Lack of a one-hand interaction method may impact the commercial success of a technology.

Evaluations are often conducted to test a refinement to existing practice. Often the new technique is an improvement over the status quo. In the next section, we present some key initiatives in improving current practice through language and movement modeling.

3. OPTIMIZATION TECHNIQUES

There are two popular approaches to optimizing the text entry task: movement minimization and language prediction. Movement minimization seeks

Figure 3. Simultaneous presentation of results for speed and accuracy. From "A comparison of three methods of character entry on pen-based computers," by I.S. MacKenzie, R. B. Nonnecke, J. C. McQueen, S. Riddersma, and M. Meltz, 1994, *Proceedings of the Human Factors Society 38th Annual Meeting.* Copyright 1994 by the Human Factors and Ergonomics Society. Reprinted with permission.

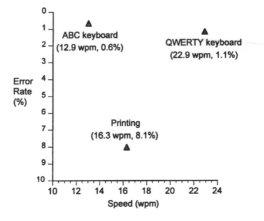

to reduce the movements of the finger or pen in interacting with a mobile device to enter text. Language prediction exploits the statistical nature of a language to predict the user's intended letters or words. There are also hybrid approaches. In the following sections we summarize these modeling and design techniques.

3.1. Movement Minimization

The main reason for using a Qwerty keyboard for text input is to support touch typing. The next-best reason is familiarity with the letter arrangement. However, the sheer size of a Qwerty keyboard is imposing and ill-suited to the mobile paradigm. Recent work has focused on the limited case of single-finger or stylus entry, either on a soft keyboard or on a small physical keyboard with a reduced key set. This work combines a statistical language model with a movement time prediction model to assist in modeling and designing input techniques wherein device or hand movement is as efficient as possible (Hunter, Zhai, & Smith, 2000; Lewis, Allard, & Hudson, 1999a; Lewis, LaLomia, & Kennedy, 1999a, 1999b; MacKenzie & Zhang, 1999; MacKenzie, Zhang, & Soukoreff, 1999; Zhai, Hunter, & Smith, 2000; Zhang, 1998). The following is the summary of a model introduced by Soukoreff and MacKenzie (1995).

The model comprises five major components: (a) a digitized layout of a keyboard; (b) Fitts' law for rapid aimed movements; (c) the Hick–Hyman law for choice selection time; (d) a linguistic table for the relative frequencies of letter pairs, or digrams, in common English; and (e) a spreadsheet or software tool in which the preceding components are combined.

For (a), each key is assigned an x–y coordinate, thus allowing digram distances to be easily computed using the Pythagorean identity. For (b), we use Fitts' law (Fitts, 1954; MacKenzie, 1992) to predict the movement time (MT; in seconds) to tap any key given any previous key. This is a simple prediction based on the distance between the keys (A_{ij}) and the size, or width, of the target key (W_j):

$$MT_{ij} = 0.204 \log_2 \left(\frac{A_{ij}}{W_j} + 1 \right) \tag{1}$$

For (c), we use the Hick–Hyman law (Hick, 1952; Hyman, 1953) to predict the reaction time (RT; in seconds) to visually scan a 27-key layout to find the target key. For novices, we set

$$RT = 0.200 \log_2 (27) = 0.951 \text{ sec.} \tag{2}$$

For experts, we set $RT = 0$ sec.

For (d), we use a 27×27 matrix of digram frequencies to establish probabilities for each digram in common English, P_{ij}. The table includes the 26 letters plus the Space character. These are used to weight the movement time predictions in obtaining the mean movement time over all possible digrams:

$$\overline{MT} = \sum_i \sum_j P_{ij} \times \left(M_{ij} + RT \right) \tag{3}$$

RT is set to either .951 seconds (novices) or 0 seconds (experts), as noted earlier.[5]

Entry speed in wpm is calculated by taking the reciprocal of the mean movement time, multiplying by 60 seconds per minute, and dividing by 5 characters per word:

$$Entry\_Speed = \left(\frac{1}{\overline{MT}} \right) \times \frac{60}{5} \tag{4}$$

---

5. A recent experiment has revealed several weaknesses in the novice component of the model (MacKenzie & Zhang, 2000). Work is underway to refine the motor component of novice model to generate more accurate predictions.

The model takes particular care to accommodate the Space bar because it is the most prevalent character in text entry tasks. The result is a general behavioral description and predictive model of the task of text entry with a stylus and soft keyboard. We consider the predictions approximate but useful (for more details, see Soukoreff & MacKenzie, 1995).

This model has been subsequently used by others seeking an optimal keyboard layout for stylus typing (Hunter et al., 2000; MacKenzie et al., 1999; Zhai et al., 2000; Zhang, 1998). Their efforts are reported later in this article.

3.2. Language Prediction

Predictive text input techniques strive to reduce the input burden by predicting what the user is entering. This is accomplished by analyzing a large collection of documents—a corpus—to establish the relative frequency of characters, digrams (pairs of characters), trigrams, words, or phrases in the language of interest. These statistical properties are used to suggest or predict letters or words as text is entered. The seminal publication in the area of text prediction is by Shannon (1951), and although there are many ways to implement text prediction, most are based on this article.

Predictive input technologies have the capacity to significantly reduce the effort required to enter text—if the prediction is good. However, there are a few caveats to consider in basing a language model on a standard corpus, including (a) the corpus may not be representative of the user language, (b) the corpus does not reflect the editing process, and (c) the corpus does not reflect input modalities. An explanation of these points follows.

Corpus Not Representative of the User Language

The idea that a corpus is "representative of a language" is questionable when the domain is users interacting with computing technology. Users typically use a much richer set of characters and words than appear in any corpus, and the statistical properties in the user's set may differ from those in the corpus. A simple example is the Space key, which is the most common character in English text (Soukoreff & MacKenzie, 1995). However, the Space character is typically missing in tables of letter or digram probabilities used to build language models (e.g., Maynzer & Tresselt, 1965; Underwood & Schulz, 1960).

In addition, punctuation symbols are rarely included in letter or digram tables. Both Isokoski (1999) and Zhai et al. (2000) observed that some punctuation symbols occur more frequently than some of the less frequent letters. Inclusion of the Space character and simple punctuation symbols is the first step. We feel it is important to fully open the character set. (Corpora often do not distinguish between capital and lowercase letters, but this is the special case of input modalities, discussed next.)

The characteristics of the text users enter are dependent on the application used to create the text. For example, we expect more formal prose using a word processor than an e-mail application. In addition, the type of application depends on the input device available—few people have the patience to enter volumes of text into a handheld PDA device. The kinds of text most likely entered in this context are short notes, phone numbers, URLs, acronyms, slang, and so forth, the statistical properties of which differ from formal English texts. Highly cryptic messages are common for text entry on cell phones (Grinter & Eldridge, in press).

Corpus Ignores the Editing Process

A corpus contains no information about the editing process, and we feel this is an unfortunate omission. Users are fallible, and the creation of a text message—or interaction with a system on a larger scale—involves much more than the perfect linear input of alphanumeric symbols. The input process is really the editing process.

Recently, we conducted a study to monitor and analyze keystroke-level interaction with desktop systems. Over a period of 2 months we logged all keystrokes (>400,000) for four desktop computer users. Figure 4 shows the 15 most common keystrokes. Common editing keys, such as Down, Back, and Up, figure very prominently in the table. Although mobile users engage a much different interface, the data in Figure 4 serve as a warning flag that input with computing technology, in general, is much richer than represented in a corpus.

Corpus Does Not Capture Input Modalities

Text documents do not reflect how they were created. For example, a corpus includes both capital and lowercase characters. In simple language models this distinction is ignored (e.g., *A* and *a* are considered the same). A more expansive model can easily accommodate this distinction simply by treating capital and lowercase characters as distinct symbols. However, from the input perspective, both approaches are wrong. Uppercase and lowercase characters are never entered via separate keys on a keyboard; thus, the seemingly more accurate treatment of uppercase and lowercase characters as distinct symbols is just as wrong.

For the user's interaction with the Shift and Caps Lock keys to be accommodated in a model of text input, activity with these and related keys should be included in the language model. In other words, it is the "language of inter-

Figure 4. Relative frequency (%) of the 15 most frequent keystrokes from four users.

S1		S2		S3		S4		All	
9.18	Space	10.42	Down	12.87	Space	8.75	Space	11.29	Space
7.14	Back	7.95	Space	8.69	Back	8.72	Back	7.10	Back
5.29	Down	5.57	Up	7.36	E	4.85	E	6.29	E
4.93	E	5.35	Shift	6.07	T	4.39	Down	5.11	T
4.19	A	5.33	Right	5.05	O	4.29	Return	4.29	O
3.85	Shift	4.49	Control	4.64	I	4.01	T	4.03	I
3.84	I	4.00	E	4.45	A	3.89	Shift	3.95	A
3.42	O	3.96	Left	4.18	S	3.88	O	3.94	Shift
3.28	T	3.73	Delete	4.16	N	3.83	I	3.78	S
3.27	R	3.25	T	3.79	R	3.57	R	3.57	N
3.22	N	3.12	S	3.46	Shift	3.31	A	3.33	R
2.98	Up	2.54	O	2.68	H	3.21	S	3.27	Down
2.92	Right	2.54	A	2.32	L	3.18	N	2.39	Delete
2.72	S	2.42	Back	2.24	C	2.84	D	2.32	H
2.48	Delete	2.38	I	2.14	D	2.26	H	2.22	C

action" that should be modeled. Note in Figure 4 that the Shift key fares no worse than 11th in the list of most-frequent keys.

3.3. Hybrid Input Techniques

Some text input techniques include both movement-minimizing and predictive features. Dasher (Ward et al., 2000) is a predictive text input technique using a pointing device to select from anticipated options (see also Ward, Blackwell, & MacKay, 2002). The options are presented to the user in boxes sized according to their relative probabilities. The boxes scroll and expand as the pointing device hovers near them (using graphics somewhat like a video game), allowing fast text entry. Thus, the technique is both movement minimizing and predictive. An online demonstration is available (http://wol.ra.phy.cam.ac.uk/mackay/dasher).

3.4. Key Minimization Techniques (Modes)

Because space is limited on small devices, keyboards that minimize the number of keys are of interest. However, users desire a large set of characters including the alphabet, numbers, symbols, and editing keys. An example of this is the standard PC-compatible 101-key keyboard. Although the standard PC keyboard has 101 keys, a user can produce closer to 800 individual keystrokes (each key is pressed in combination with Shift, Ctrl, or Alt, and the

Num Lock key changes the mode of the numeric keypad). The keys on the standard PC keyboard are, therefore, ambiguous; disambiguation is accomplished with the various mode keys.

There is another way to disambiguate keystrokes. Some keyboards are designed with more than one letter on each key (e.g., the alphabetic characters on a standard telephone keypad). Text entered on these is inherently ambiguous because different character strings correspond to the same key presses. For example, on a standard telephone keypad, both *gap* and *has* correspond to the key sequence 4–2–7. Disambiguation technology takes key press sequences and uses an embedded database of language statistics to identify legal words. These are presented to the user for verification. Automated disambiguation holds promise to increase the speed and accuracy of text input on ambiguous keyboards.

Conceptually, we can think of key ambiguity as a continuum (see Figure 5). At one extreme, we have a keyboard with a dedicated key for each symbol in the language (Figure 5a), whereas at the other extreme we have just one key that maps to every symbol in the language (Figure 5d). The keyboard in Figure 5d would be very fast[6] because only one key is pressed. However, it is of no practical use because each key press is ambiguous to the entire set of symbols in the language. Clearly, Figure 5d is little more than a curiosity. The Qwerty keyboard (Figure 5b) and telephone keypad (Figure 5c) represent two relevant points in the continuum.

The previous sections introduced many issues facing researchers in mobile text input, and we have delineated the design space within which this research takes place. In the following section we present a survey of mobile text entry techniques as found in research papers and commercial products.

4. SURVEY OF TEXT ENTRY TECHNIQUES

The survey is divided into key-based and stylus-based text input methods.

4.1. Key-Based Text Entry

Key-based text entry techniques range from those that use a keyboard where each key represents one or more letters to those with as few as three keys.

---

6. In fact, the text entry rate for this keyboard would be about 78.4 wpm. This figure is derived from the single finger key repeat time of .153 sec reported by Soukoreff and MacKenzie (1995). The text entry rate is $(1 / .153) (60 / 5) = 78.4$ wpm. The key repeat time may be as low as .127 sec (Zhai, Hunter, & Smith, 2000), and in this case, the upper bound is $(1 / .127) (60 / 5) = 94.5$ wpm.

Figure 5. The key-ambiguity continuum: (a) fictitious alphabetic keyboard with distinct keys for capital and lowercase letters; (b) Qwerty keyboard; (c) standard telephone keypad; (d) hypothetical single-key keyboard, which, to be useful, would require either many mode keys or a near psychic disambiguation algorithm.

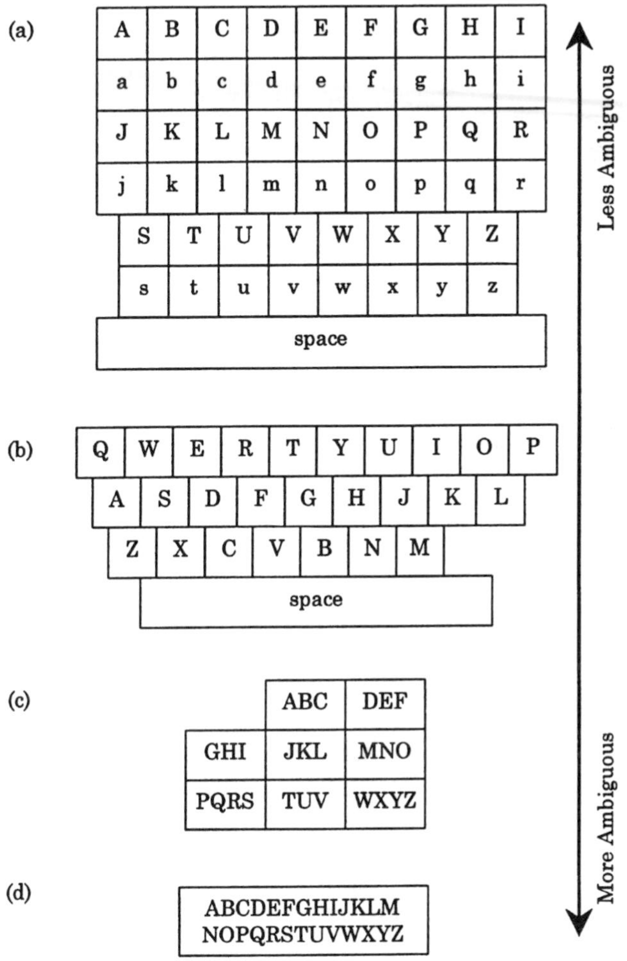

Telephone Keypad

The desire for an effective text entry method using the telephone keypad is fueled by the increase in text-messaging services and the movement toward consolidation of technologies such as wireless telephony and handheld computers. Text entry on a mobile phone is based on the standard 12-key telephone keypad (see Figure 6).

Figure 6. The standard 12-key telephone keypad.

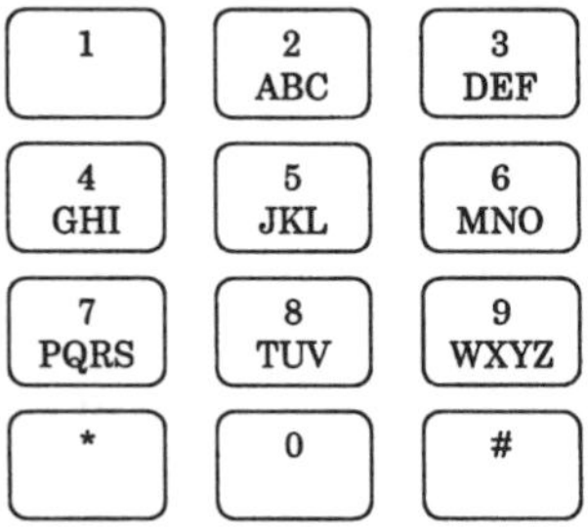

1	2 ABC	3 DEF
4 GHI	5 JKL	6 MNO
7 PQRS	8 TUV	9 WXYZ
*	0	#

The 12-key keypad consists of number keys 0 to 9 and two additional keys (* and #). Characters A to Z are spread over keys 2 to 9 in alphabetic order. The placement of characters is similar in most mobile phones, as it is based on an international standard (Grover, King, & Kuschler, 1998). The placement of the Space character varies among phones; however, it is usually entered with a single press of the 0 key or the # key. Because there are fewer keys than the 26 needed for the characters A to Z, three or four characters are grouped on each key, and so, ambiguity arises, as noted earlier. In the following paragraphs, we present three approaches to text entry on a phone keypad: multitap, two-key, and one-key with disambiguation.

The multitap method is currently the most common text input method for mobile phones. With this approach, the user presses each key one or more times to specify the input character. For example, the 2 key is pressed once for the character A, twice for B, and three times for C. Multitap suffers from the problem of segmentation, when a character is on the same key as the previous character (e.g., the word *on* because both O and N are on the 6 key). To enter the word *on*, the user presses the 6 key three times, waits for the system to time-out, and then presses the 6 key twice more to enter the N. Another segmentation technique is to use a special key to skip the timeout ("timeout kill"), thus allowing direct entry of the next character on the same key. Some phone models use a combination of the two solutions. For example, Nokia phones (Nokia Group, Finland; http://www.nokia.com) include both a 1.5-sec timeout and the provision for a timeout kill using the Down Arrow key. The user decides which strategy to use.

In the two-key method, the user presses two keys successively to specify a character. The first key selects the group of characters (e.g., the 5 key for J, K, or L). The second key specifies the position within the group. For example, to enter the character K the user presses 5 followed by 2 (K is second character in JKL). Although the two-key method is quite simple, it is not in common use for entering Roman letters. However, in Japan a similar method (often called the "pager" input method) is very common for entering Katakana characters.

A third way to overcome the problem of ambiguity is to add linguistic knowledge to the system. We call this technique *one-key with disambiguation.* An example is T9® by Tegic Communications, Inc. (Seattle, WA; http://www.tegic.com). When using T9 each key is pressed only once. For example, to enter *the*, the user enters 8–4–3–0. The 0 key for "space" delimits words and terminates disambiguation of the preceding keys. T9 compares the word possibilities to a linguistic database to guess the intended word.

Naturally, linguistic disambiguation is not perfect because multiple words may have the same key sequence. In these cases the most common word is the default. A simple example follows using the well-known "quick brown fox" phrase: (words are shown top to bottom, most probable at the top)

843	78425	27696	369	58677	6837	843	5299	364
the	**quick**	**brown**	**fox**	**jumps**	**over**	**the**	jazz	**dog**
tie	stick	crown		lumps	muds	tie	**lazy**	fog
vie							vie	

Of the nine words in the phrase, eight are ambiguous, given the required key sequence. For seven of the eight, however, the intended word is the most probable word. The intended word is not the most probable word just once, with *jazz* being more probable in English than *lazy*. In this case, the user must press additional keys to obtain the desired word. Evidently, the term *one-key* in "one-key with disambiguation" is an oversimplification!

Silfverberg, MacKenzie, and Korhonen (2000) presented predictive models of these three text input methods based on the model of Soukoreff and MacKenzie (1995). They reported that the disambiguation of T9 works reasonably well, with expert predictions ranging from 41 to 46 wpm. However, these figures are coincident with rather broad assumptions. These include (a) all words entered are unambiguous, (b) users are experts (i.e., no typing, spelling, or other errors), and (c) all words entered are in the dictionary. Their predictions are, at best, an upper bound.

Many mobile phone manufacturers have licensed the T9 input technology, and since 1999 it has surfaced in commercial products (e.g., the Mitsubishi MA125, the Motorola i1000Plus, and the Nokia 7110). There is also a touch screen version of T9 that is available for PDAs. Bohan, Phipps, Chaparro, and Halcomb (1999) described an evaluation of the touch screen version.

T9 was the first disambiguating technology to work with a standard mobile phone keypad, but not the only such technology. Motorola's iTAP® is disambiguating technology similar to T9. Both iTAP and T9 support multiple languages. The Chinese version of iTAP uses a nine-key input method for writing the various strokes; it offers users more keystroke choices and is easy to learn (Sacher, 1998). Another similar technology is eZiText® by Zi Corporation

(Calgary, Alberta, Canada; http://www.zicorp.com). No published evaluations exist of iTAP or of eZiText.

A slightly different approach is presented in WordWise by Eatoni Ergonomics (New York; http://www.eatoni.com). To aid in disambiguation, a mode shift is used to explicitly choose one character from each key and the other characters remain ambiguous; this achieves partial disambiguation. Figure 7 illustrates the WordWise keypad.

The mode shift is implemented either with the 1 key (shown in Figure 7) or using a thumb-activated key on the side of the mobile phone. To enter the letter C, the Shift key is pressed followed by the 2 key. To enter the letter A, the 2 key is pressed by itself, and automatic disambiguation determines whether the user intended to enter A or B. The letters chosen for the mode shift are C, E, H, L, N, S, T, and Y, most of which are the most popular letters in each group (on each key). These letters were chosen to provide maximum separation for the disambiguation algorithm. One beneficial side effect of the mode shift is that words that are explicit (e.g., *the*, which is entered by holding Shift while entering 8–4–3) can be omitted in the internal database. This greatly reduces the memory requirements of the implementation—a critical factor for mobile phones.

Text input on the telephone keypad, working in concert with language-based disambiguation (e.g., T9 or WordWise), requires the attention of the user to monitor the outcome of keystrokes. A typical text creation task has two FOA because the user attends to both the keypad and the display. The performance impact of this behavior is difficult to model because it depends on cognitive and perceptual processes, and on user strategies (see Silfverberg et al., 2000, for further discussion).

With the multitap or two-key techniques, the outcome of keystrokes bears no such uncertainty; thus, skill in performing eyes-free input is more easily attained. The models created by Silfverberg et al. (2000) predict about 21 to 27 wpm for the multitap method and the two-key method.

Small Qwerty Keyboards

The most prevalent text input technology for low-end PDAs is the miniature Qwerty keyboard. There are many examples, such as the HP2000, some models of the HP Jornada, the Sharp (Osaka, Japan; http://sharp- world.com) Zaurus, the Sharp Mobilon, and the Psion (London; http://psion.com) Revo. Two-way pagers support text input and at least two companies have pager products with miniature Qwerty keyboards.

The BlackBerry by Research In Motion is a two-way pager with a small Qwerty keyboard (see Figure 8a). The keyboard is too small for touch typing, but it is suitable for one- or two-finger typing. Motorola has a similar product called the PageWriter (see Figure 8b).

Figure 7. Eatoni Ergonomics WordWise keypad. The 1 key acts as a shift to explicitly select one letter on each alpha key.

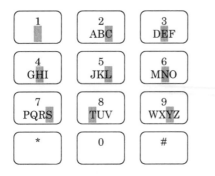

Figure 8. (a) Research in Motion BlackBerry (RIM 957; actual size 79 × 117 mm) and (b) Motorola PageWriter 2000X (actual size 95 × 71 mm).

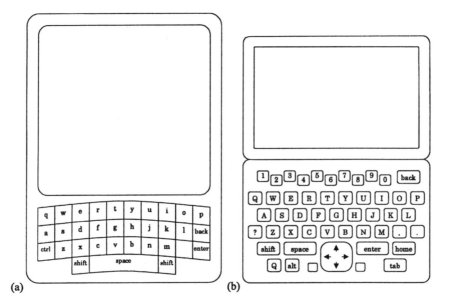

The Nokia Communicator is a mobile phone with text-messaging function-ality. It looks like a typical mobile phone when operated as a phone, but it opens to reveal a large LCD screen and miniature Qwerty keyboard inside (see Figure 9).

The BlackBerry, PageWriter, and Communicator are representative of small devices that have stayed with the Qwerty paradigm, and they are by no means alone. There are many similar devices on the market.

Figure 9. Nokia 9110 Communicator (actual size 158 × 112 mm).

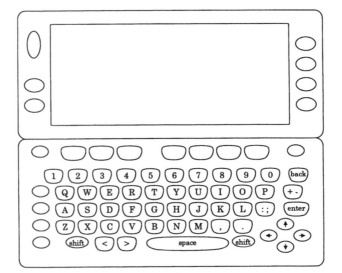

There is another way to reduce the size of a Qwerty-like keyboard. Matias and colleagues proposed a clever way to half the size of the keyboard and still leverage touch typing skills (Matias, MacKenzie, & Buxton, 1993, 1994, 1996a, 1996b). The Half-Qwerty keyboard, commercialized by Matias Corporation (Rexdale, Ontario, Canada; http://www.halfqwerty.com), is a regular Qwerty keyboard that is split in half. There are two possible Half-Qwerty keyboards—one corresponds to the left half of the Qwerty keyboard and the other to the right. To enter characters the user simply types the appropriate key in the regular fashion, but if the Space bar is held while a key is typed, the corresponding character from the other half of the keyboard is entered. Either hand can be used. Hitting the Space bar alone types a space. Note that the relative finger movements used for one-handed typing are the same as those used for two-handed typing. The two Half-Qwerty keyboards are depicted in Figure 10.

Matias and colleagues report the results of a rigorous user evaluation of the Half-Qwerty (Matias et al., 1993, 1996a). Right-handed participants using their left hands reached 50% of their two-handed typing speed after approximately 8 hr of practice, and after 10 hr all participants typed between 41% and 73% of their two-handed speed, ranging from 24 to 43 wpm.

The Half-Qwerty keyboard is unique among the other solutions to the mobile and handheld text entry problem because the keyboard is small, familiar to users, supports fairly rapid text entry, and has some significant applications. There are many industrial jobs that require a worker to enter text with one

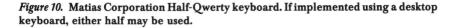

Figure 10. Matias Corporation Half-Qwerty keyboard. If implemented using a desktop keyboard, either half may be used.

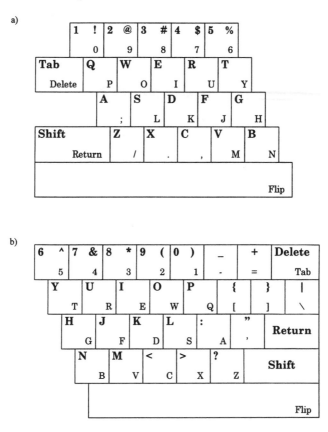

hand while doing another task with the other. The Half-Qwerty keyboard is also useful in situations where a user has lost the use of one hand. In both cases, software can be installed on a regular desktop computer that enables Half-Qwerty functionality. A small stand-alone version is now available as an add-on for handheld devices.

Although lugging around a full-sized Qwerty keyboard to use with the PDA in one's shirt pocket seems odd, collapsible Qwerty keyboards allow users to do just that. In 1999 Think Outside Inc. (Carlsbad, CA; **http://www.think-outside.com**) released the Stowaway™, a full-size Qwerty keyboard that collapses to a 91 × 130 × 20 mm volume. Originally released for the Palm, the Stowaway was later adopted by Palm Computing, becoming the Palm™ portable keyboard. Think Outside also produce collapsible keyboards for other families of PDA.

Five-Key Text Entry

By way of introduction to five-key text input, we mention the *date stamp* method (also known as the *three-key* method). This method can be implemented using very limited hardware: technology to display at least one character, two buttons (or a wheel) to scroll through the alphabet, and an Enter key. It is called the date stamp method because, similar to a date stamp, the desired character is selected by rotating through the character set. Video arcade games often use this technique for players to enter their name when they achieve a high score. The technique is also commonly used for entering text into some electronic musical instruments. Although the three-key method is reasonable for entering small amounts of text into devices with a simple interface, the method is frustratingly slow and not suitable for even modest amounts of text entry (see Bellman & MacKenzie, 1998, for further discussion).

Five-key text entry uses an interface with four cursor keys (up, down, left, and right) and an Enter key (see Figure 11). The alphabet, number, and symbol characters are presented on a LCD display with typically three to five rows and 10 to 20 columns, and the five keys are used to move a cursor and select one letter at a time.

The characters are presented in alphabetic order or in the familiar Qwerty arrangement. The five-key input method is typically used on very small devices like the recent generation of pagers, which only have enough space for a small LCD screen and five keys. An example is the AccessLink® II pager from Glenayre Electronics Inc. (Charlotte, NC; http://www.glenayre.net).

The main problem with the five-key method is that many key presses are required to move between characters, and this significantly slows input. In view of this, Bellman and MacKenzie (1998) devised a technique known as *fluctuating optimal character layout* (FOCL). The idea is that because the input device knows the last character the user has entered, it can subsequently present the characters in an arrangement that places the most likely characters closer to the cursor's home position; the display is rearranged after each character entered so as to minimize the number of cursor movements to select the most likely next character. They show that the average number of KSPC can be reduced by over 50%, from just over 4 KSPC for the alpha layout to less than 2 KSPC using FOCL.

Bellman and MacKenzie (1998) reported the results of an exploratory study comparing FOCL to the five-key input method using the Qwerty arrangement of letters. Their study, with 10 participants, found that after 10 sessions of 15 min each there was no statistically significant difference in entry speed or accuracy. The average speed they reported for both Qwerty and FOCL is 10 wpm. Although the study was longitudinal in nature, evidently participants did not have enough exposure to FOCL to approach their maximum text entry

Figure 11. Five-key text entry.

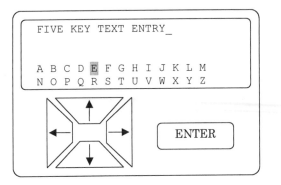

speeds. Although fewer keystrokes were required to enter each character, more visual scan time was required to find the next character. In short, as with many other optimized text entry methods, the advantage of the input technique is apparently not realized until users invest considerable time to become familiar with the new technology.

The three-key, five-key, and FOCL text input techniques all require the user's attention on the screen to scroll around and select characters. Therefore, text creation is a two-FOA task with these techniques. Single-handed input is possible with all of these input techniques.

Other Small Keyboards

Some researchers have proposed alternatives to the telephone or Qwerty key arrangements. The *single-hand key card* (SHK) is a small card with a keyboard and joystick proposed by Sugimoto and Takahashi (1996). The SHK is held in one hand, pinned between the palm and the thumb in such a way that the four fingers manipulate the keyboard and joystick on the top face of the device. SHK is a small keyboard with multiple characters on each key. It employs disambiguation technology. The keyboard arrangement of SHK appears in Figure 12. The joystick and three function keys appear in a row above the keyboard on the device (not shown). The AR key in Figure 12 toggles through the word possibilities generated by the ambiguity resolution feature.

Sugimoto and Takahashi (1996) reported that the keys were arranged so as to reduce the average motion of the fingers, although they have not explained in detail how they came to their key arrangement or published an evaluation of their device. Once the arrangement is learned by the user, the device could support single FOA text creation and single-handed text input.

Another important class of keyboard is *chording keyboards*, where text is entered by pressing multiple keys simultaneously. Because multiple keys are

Figure 12. The key arrangement of the Single-Hand Key card device.

P	G	C	Z	W	Shift
	N	T	R	K	J	
H	S					
A	E	I	O	AR	Enter
U	X	Y	V	Q	Space
	D	F	M	L	B	

pressed, fewer keys are needed on a chord keyboard (resulting in smaller devices) and chords not being used for entering single letters can be used to enter words. The Twiddler by Handykey Corporation (Denver, CO; http://www.handykey.com) is a chord keyboard popular with researchers in the wearable and ubiquitous computing fields. The Twiddler is operated with one hand and has 4 mode keys depressed by the thumb (Number, Alt, Ctrl, and Shift) and 12 keys for the fingers. The Twiddler keyboard appears in Figure 13; notice that the layout of the characters is somewhat alphabetical. However, the Twiddler is user configurable; the user may change the characters (or words) entered by each chord (the keyboard appearing in Figure 13 is the default layout), and other character mappings for the chords have been proposed for the Twiddler, which are claimed to map common characters to easier chords. The Twiddler also has chords defined for common small English words and parts of words (e.g., *the, and, -ion,* and *-ing,* etc.).

The Twiddler is by no means the only chord keyboard. Other examples include the BAT™ by Infogrip, Inc. (Ventura, CA; http://www.infogrip.com) and MonoManus® by ElmEntry Enterprises (Minneapolis, MN; http://www.hankes.com/eee/index.htm). However, most of these keyboards interface to desktop computers and are not expressly for mobile computing platforms. The Twiddler interfaces to the Palm only if the Happy Hacking™ Cradle (by PFU America, Inc; San Jose, CA; http://www.pfuca.com) is used (available separately).

The Twiddler can be used with one hand (zero FOA, once the chords are learned), and anecdotal reports of typing speed as fast as 50 wpm have been reported (Hjelm, Tan, Fabry, Fanchon, & Reichert, 1996).

4.2. Stylus-Based Text Entry

Stylus-based text entry uses a pointing device, typically a pen (a.k.a. stylus), to select characters through tapping or gesture. Although our discussions here are limited to stylus input, there are several related examples of research in

Figure 13. The key arrangement of the Twiddler chord keyboard device. Letters with a white background are entered by pressing the key by itself. Letters with a light grey background are entered by pressing the key and the E key simultaneously. Letters with a dark grey background are entered by pressing the key and the A key simultaneously.

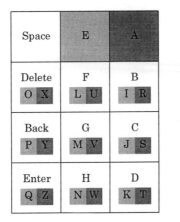

mobile text entry using finger or touch input, wherein the user's finger is used instead of a stylus (e.g., Enns & MacKenzie, 1998; Fukumoto & Suenaga, 1994; Goldstein, Book, Alsio, & Tessa, 1999). All of the stylus-based text entry techniques require two hands, unless the user can support the device on a table while using it.

Traditional Handwriting Recognition

Handwriting recognition was once touted as *the* solution for mobile text entry, but early systems received considerable bad press, as noted earlier. To be fair, handwriting recognition is a difficult problem, and the technology has improved since the early days. There are two problems that handwriting recognizers must solve: segmentation and recognition. The input to a recognizer is a series of ink trails, with each stored as a set of digitized points representing the stylus travel between pen-down and pen-up actions. Segmentation is the process of determining which segments are in which characters. With the goal of supporting "natural handwriting," input is often a mixture of block printing and cursive handwriting. As one might imagine, segmenting the strokes in the sloppy scrawl of a user is very difficult indeed. One way to reduce the complexity is to constrain input (e.g., to support block printed characters only). However, entry like this is by no means "natural." Generally, the more relaxed the constraints, the more difficult the segmentation and recognition process; recognition accuracy usually suffers. To compensate, recognizers are made more complex and, unfortunately, require more memory (see Tappert, Suen, & Wakahara, 1990, for a detailed survey of recognition techniques and technologies).

One obstacle for recognition-based technologies is high user expectations. LaLomia (1994) reported that users are willing to accept a recognition error rate of only 3% (a 97% recognition rate), although Frankish and colleagues (1995) concluded that users will accept higher error rates depending on the text-editing task. Several researchers have published studies evaluating or comparing the recognition rate of various recognition systems. Chang and MacKenzie reported a recognition rate of 87% to 93% for two recognizers (Chang & MacKenzie, 1994; MacKenzie & Chang, 1999). Wolf, Glasser, and Fujisaki (1991) reported a recognition rate of 88% to 93%. Santos, Baltzer, Badre, Henneman, and Miller (1992) reported a novice recognition rate of 57%, although this improved to 97% after 3 hr of practice. These studies suggest that recognition technology is close to matching user expectations for expert users but that novices may be discouraged by their initial experiences. Perhaps the acid test, an observation suggesting that handwriting recognition does not yet perform adequately, is that there are no mobile consumer products in the market today where natural handwriting recognition is the sole text input method. The products that do support stylus-based text input work with constraints or stylized alphabets (see later).

Gibbs (1993) made an important observation on text entry speed and handwriting recognition. In his summary of 13 recognizers, the recognition speed of the systems was at least 4 cps, which translates into 48 wpm. However, human hand-printing speed is typically on the order of 15 wpm (Card et al., 1983; Devoe, 1967; Van Cott & Kinkade, 1972). In other words, speed is a function of human limitations, not machine limitations. Even with perfect recognition, therefore, entry rates can never reach those of, for example, touch typing.

Unistrokes

Unistrokes is a stylized single-stroke alphabet developed by Goldberg and Richardson at the Xerox Palo Alto Research Center (Goldberg & Richardson, 1993). At the time of the invention, handwriting recognition technology was not in good stead with users, as the problems noted earlier were rampant in existing products. To address these Goldberg and Richardson developed a simplified set of strokes that is both easier for software to recognize and quicker for users to write. The Unistrokes alphabet appears in Figure 14.

The name *Unistrokes* describes the most significant simplification that Goldberg and Richardson (1993) made: Each letter is written with a single stroke. This greatly simplifies recognition, as the segmentation problem is essentially eliminated. The strokes are so simple that users can write Unistrokes

Figure 14. Unistrokes alphabet.

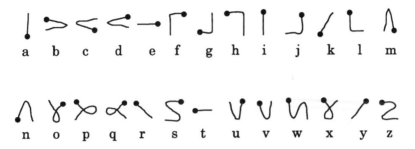

without watching the stylus. Goldberg and Richardson observed that Unistrokes afford what they termed *heads-up text entry* (i.e., reduced FOA). The Unistrokes alphabet does not contain numbers, punctuation, or symbolic characters, although the original publication (Goldberg & Richardson, 1993) suggests ways of supporting these (e.g., using a dedicated stroke as a mode shift).

Although a comparative study of Unistrokes has never been undertaken, some experimental results are given (Goldberg & Richardson, 1993). Ignoring errors, a text-entry rate of 2.8 cps (i.e., 34 wpm) was reported.

Although an interesting and promising idea, Unistrokes did not catch on, and the most likely reason is that the strokes are not similar enough to regular handwritten or printed letters; the strokes must be learned. Palm designed a single-stroke system called Graffiti that is used in their Palm product. Graffiti has been credited as a significant reason for the commercial success of the Palm (Blickenstorfer, 1995). The Graffiti alphabet appears in Figure 15.

Graffiti has strokes for punctuation, numbers, symbolic characters, and mode switches (capital vs. lowercase). These are omitted in Figure 15 for brevity. Capital and lowercase characters are supported with mode switching, which is accomplished with a dedicated stroke. Basic editing (Backspace) is also supported with a special stroke.

The great advantage that Graffiti has over Unistrokes is its similarity to normal hand-printed characters. MacKenzie and Zhang (1997) performed a study of the immediate usability of Graffiti. They observed that 79% of the Graffiti strokes match letters of the Roman alphabet. Under experimental conditions they measured the accuracy with which participants could enter the alphabet following 1 min of studying the Graffiti reference chart, following 5 min of practicing with Graffiti and following a 1-week lapse with no intervening practice. The accuracies they reported were very high–86%, 97%, and 97%, respectively.

Isokoski (1999) presented a single-stroke alphabet that can be entered using a wide range of pointing devices. He observed that the easiest motions to make with pointing devices are the four primary compass directions: up, down, left, and right. Another design objective was finding the optimal mapping between

Figure 15. Graffiti alphabet.

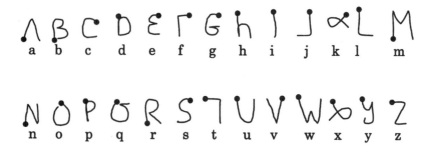

the four directional strokes and the characters of the alphabet (more frequent characters should have shorter strokes). He called the result *minimal device-independent text-input method* (MDTIM). The MDTIM alphabet appears in Figure 16.

Isokoski (1999) evaluated his single-stroke alphabet with a variety of pointing devices. The measured average text entry speed using a touch pad was 7.5 wpm. The study was not longitudinal, and the participants were still showing improvement at the end of the trials. The MDTIM alphabet suffers from the same affliction as Unistrokes and many other text input methods: The alphabet is not familiar to the average user, and practice is required to learn the alphabet and attain fast entry speeds. However, Isokoski's results do indicate that the MDTIM alphabet is indeed device-independent.

Until recently Windows CE devices were without a similar easy-to-learn handwriting recognition technology. This changed in 1998 when Microsoft licensed Jot from Communication Intelligence Corporation. Jot recognizes many of the Graffiti strokes and a number of alternative strokes similar to normal handwriting and printing as well. The Jot alphabet appears in Figure 17.

Jot also includes strokes for numbers, symbolic characters, and common editing functions. The different cases (capital vs. lowercase) are selected by where the user writes the stroke on the touch screen of the device. Jot also allows some customization: Users can indicate writing preferences for some characters.

All the alphabets just described have the potential to support single-FOA text entry once the user is familiar with the stylized alphabet.

Gesture-Based Text Input

Gestures are informal motions for communication. We classify the text entry methods in this section as gestural because of their informality and fluidity. Character-recognition-based and soft-keyboard-based input techniques have fixed characters that are entered in a certain way, or the stylus must be tapped in a certain location to select characters for input. Gesture-based text input

Figure 16. MDTIM alphabet.

Figure 17. Jot alphabet.

technologies do not have a fixed set of strokes that a recognizer turns into characters; gestural text input methods have a framework in which informal stylus motions are interpreted as characters.

An example of this is Cirrin, a technology presented by Mankoff and Abowd (1998). The letters of the alphabet are arranged inside the perimeter of an annulus. Figure 18 shows the word *cirrin* written on the Cirrin interface.

Figure 18. The Cirrin interface, indicating how to enter the word *cirrin.*

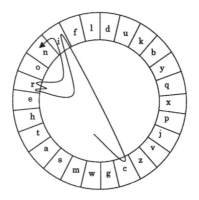

Characters are selected by moving into and out of the appropriate sector of the annulus. Mankoff and Abowd chose the circular arrangement and the order of the letters to minimize the distance between likely consecutive characters. In Figure 18, notice how the final two letters, *in,* are selected; the stylus can be moved directly from one letter into the neighboring letter.

Cirrin is not a "heads-up" text input method; users must attend to the interface when entering text. As presented, only alphabetic characters are supported. A space is entered by lifting the stylus and punctuation and mode shifts are accomplished by using an auxiliary technique, such as keys operated by the nondominant hand.

Mankoff and Abowd (1998) did not report a user evaluation in their publication; however, they stated that Cirrin "is about as fast as existing pen entry systems" (p. 214), but no indication is given of specifically what pen entry systems they compared Cirrin to.

Quikwriting is an input technology described by Perlin (1998). The idea is to have a 3 × 3 grid where characters are entered with strokes that begin in the center "home" position and move through one to three adjoining positions, returning back to the home position. Figure 19 illustrates the Quikwriting lowercase menu. Quikwriting has similar displays and modes for numbers, capitals, and symbols. The symbols in the top center and bottom center positions represent the different modes. Letters that occur more frequently in English are given the shortest strokes. For example, *i* in Figure 19 is selected by moving into the bottom right position and then returning back to the home position. Infrequent letters have longer strokes (*k* requires a move into the upper left position, then to the upper right position, and finally back to home). There is an online demonstration available at **http://www.mrl.nyu.edu/projects/ quikwriting/**. Quikwriting, like Cirrin, requires the user to look at the interface and so is a two-FOA interface, if users correct errors as they go.

Figure 19. Quikwriting lowercase interface indicating how to enter the word *quik.*

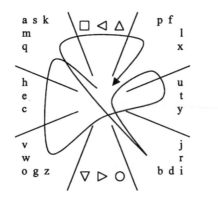

At the time of Perlin's (1998) publication, a user evaluation had not been performed, although he wrote that users familiar with Graffiti found Quikwriting about three times faster.

Another gestural text input technology is T-Cube, described by Venolia and Neiberg (1994). T-Cube is similar to a two-tier pie menu system. Figure 20 shows the pie menu structure of T-Cube. The user places the stylus within one of nine starting positions (arranged in a 3 × 3 grid; see Figure 20a). The location where the stylus is first placed indicates which of the pie menus (see Figure 20b) the user will select a character from. One of the eight characters in the submenu is chosen by flicking the stylus in the appropriate direction. The interface does not display the pie menus (Figure 20b) unless the user hesitates. T-Cube includes numbers, many symbol characters, and basic backspace editing. Like the other gestural input techniques, T-Cube requires the attention of the user, making standard text entry a two-FOA task.

Venolia (1994) presented the results of a user study of T-Cube indicating that reasonably fast text entry can be achieved; one of his participants achieved a rate of 106 characters per minute (21 wpm). However, he also acknowledged that the interface is difficult to learn.

Soft Keyboards

A soft keyboard is a keyboard implemented on a display with built-in digitizing technology. Text entry is performed by tapping on keys with a stylus or finger. However, eyes-free entry is not possible. The advantages of soft keyboards include simplicity and efficient use of space. When no text entry is occurring, the soft keyboard disappears, thus freeing screen space for other purposes.

Soft keyboards have performance advantages too. MacKenzie, Nonnecke, Riddersma, et al. (1994) reported a text entry task comparing a Qwerty soft

Figure 20. T-Cube pie menu structure: (a) first-level menu; (b) second-level menus.

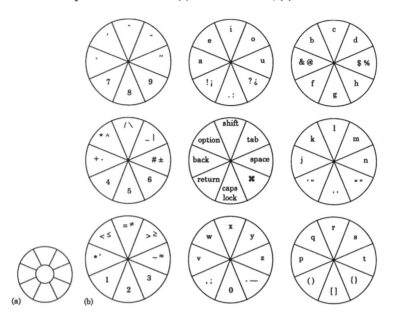

keyboard, an ABC soft keyboard, and hand printing. The Qwerty soft keyboard was both faster and more accurate than hand printing (see Figure 3).

This section presents some variations of soft keyboards developed in industry and research labs. We begin by giving the predicted expert entry rates according to the model of Soukoreff and MacKenzie (1995) presented earlier. These are given in Figure 21 sorted by predicted entry rate, highest to lowest. Several changes and one correction[7] have been introduced to the model since

---

7. The model works by using digrams to model the users' transitions from key to key as they enter text. However, a long Space key (such as in the Qwerty keyboard) or multiple Space keys are best modeled with trigrams. The error made by MacKenzie and Zhang (MacKenzie & Zhang, 1999; Zhang, 1998) was a miscalculation involving the relative probabilities of trigrams containing a Space character. Typically, trigram frequencies are not explicitly represented but, rather, are derived from digram frequencies. The probability of a trigram (i.e., the probability of the character sequence *i–j–k*) is found with the expression:

$$P(i,j,k) = P(i,j)\frac{P(j,k)}{\sum_s P(j,s)},$$

where *P(i,j)* is the probability of digram *i–j*. MacKenzie and Zhang omitted the denominator from their calculations. This error was first reported by Hunter, Zhai, and Smith (2000) and Zhai, Hunter, and Smith (2000).

Figure 21. **Expert predictions for various soft keyboard layouts.**

Keyboard Layout	Expert Prediction (wpm)	Improvement (%) Over Qwerty	Figure
Metropolis II	42.94	42.9	Figure 26c
OPTI II	42.37	41.0	Figure 25b
OPTI I	42.16	40.3	Figure 25a
Metropolis I	42.15	40.3	Figure 26b
Fitaly	41.96	39.7	Figure 23
Hook's	41.15	37.0	Figure 26a
Cubon	37.02	23.2	Figure 24
Lewis	34.65	15.3	Figure 27
ABC III	32.50	8.2	Figure 22c
ABC IV	30.18	0.5	Figure 22d
ABC II	30.13	0.3	Figure 22b
Qwerty	30.04	—	Figure 5b
DotNote	29.46	−1.9	Figure 28
ABC I	28.79	−4.2	Figure 22a

it was introduced in 1995. The entries in the table are updated from earlier publications to reflect these changes (see also Zhai et al., 2000, and elsewhere in this issue of *Human–Computer Interaction*) for a discussion of the model's sensitivity to factors such as the coefficients in the Fitts' law model and the corpus used in building the language model.

Figure 21 also gives the improvement of each soft keyboard, relative to Qwerty. At the top of the list is the Metropolis II keyboard, with a predicted text entry rate 42.9% higher than Qwerty. We visit this shortly.

There are two keyboard arrangements generally familiar to most users: Qwerty and alphabetic. The Qwerty keyboard was shown earlier (see in Figure 5b). A few alphabetic arrangements appear in Figure 22. An experiment reported by MacKenzie et al. (1999) found that participants achieved rates of 20 wpm on a Qwerty soft keyboard and 11 wpm using an ABC layout (ABC I in Figure 22a). The predicted expert entry rates are 30.04 wpm for a Qwerty soft keyboard and 28.79 wpm for the ABC I arrangement. Predictions for the ABC II, ABC III, and ABC IV arrangements are 30.13 wpm, 32.50 wpm, and 30.18 wpm, respectively (see Figure 21).

The inventors of the FITALY keyboard by Textware™ Solutions Inc. (Burlington, MA; http://www.textwaresolutions.com) used an ad hoc optimization approach to minimize the distance between common character pairs. The resulting keyboard (see Figure 23) contains two Space bars and the letters are arranged so that common pairs of letters are often on neighboring keys. MacKenzie et al. (1999) reported a walk-up (i.e., participants did not have previous experience and did not get much practice) typing rate for the

Figure 22. **Some alphabetic keyboard arrangements (a) ABC I; (b) ABC II; (c) ABC III; (d) ABC IV.**

a)
space
a b
c d
e f
g h
i j
k l
m n
o p
q r
s t
u v
w x
z y
b)
space
a n
b o
c p
d q
e r
f s
g t
h u
i v
j w
k x
l y
m z
c)
a b c d e f
g h i j k l
m n o p q r
s t u v w x
z y space
d)
a b c d e f g h i j k l m
n o p q r s t u v w x z y
space

Figure 23. **FITALY keyboard.**

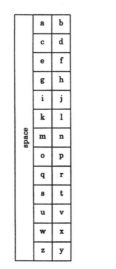

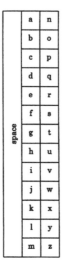

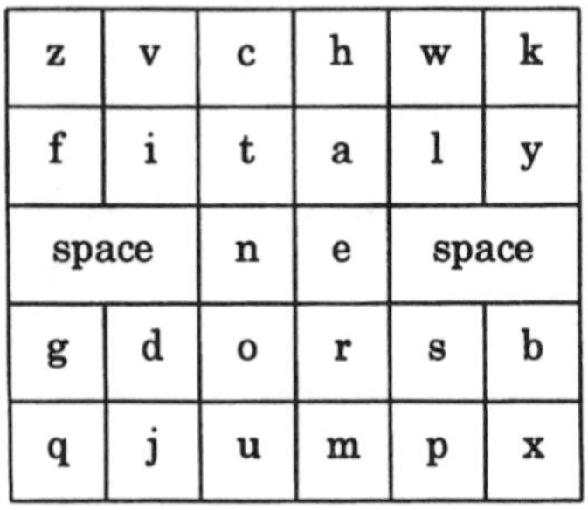

z v c h w k
f i t a l y
space n e space
g d o r s b
q j u m p x

FITALY keyboard of 8 wpm. The expert prediction for the FITALY layout is 41.96 wpm (see Figure 21).

Little information is available on the Cubon keyboard, except that it seems to have been proposed by R. A. Cubon and is used in rehabilitation situations for persons with the use of only one finger, or with a head-mounted pointing device. We know of no published user studies. The Cubon keyboard arrangement that appears in Figure 24 is given in Zhai et al. (2000). The expert prediction for Cubon is 37.02 wpm.

MacKenzie and Zhang (1999) used Soukoreff and MacKenzie's model to produce an optimized keyboard arrangement. The OPTI I keyboard appears in Figure 25a. They reported a predicted expert typing rate of 58 wpm; however, this prediction includes the error pointed out by Zhai and colleagues (see Footnote 7). Our current prediction for the OPTI I layout stands at 42.16 wpm.

MacKenzie and Zhang performed a longitudinal study over 20 sessions comparing the Qwerty and OPTI I arrangements and found that the average typing rate for OPTI I increased from 17 wpm initially to 44 wpm after 8 hr of practice (see Figure 1). For the Qwerty layout, rates increased from 28 wpm to 40 wpm over the same interval. The average rates for OPTI I exceeded those for the Qwerty layout after about 4 hr of practice.

The alert reader will notice that something is amiss: The *observed* rates actually exceeded the expert predictions! The most likely explanation is that the slope coefficient in the Fitts' law prediction model is too conservative. The slope coefficient used in the predictions is .204 sec per bit (see Equation 1), a value obtained from a pointing device study using a stylus on a Wacom tablet in a serial tapping task (MacKenzie, Sellen, & Buxton, 1991). The reciprocal of the slope coefficient is commonly known as the Fitts' law bandwidth and, in this case, is 1 / .204 = 4.9 bits per second. A discrepancy of even 1 bit per second is enough to raise the predicted rate above the observed rates.[8] Although the model is clearly sensitive to the slope coefficient in Fitts' law, adjustments do not change the rank order of the predictions in Figure 21. The reader is directed to Zhai et al. (2000) and elsewhere in this issue of *Human–Computer Interaction* for further discussion.

---

8. The bandwidth coefficient in MacKenzie, Sellen, and Buxton (1991) was measured in an "indirect" task: Participants manipulated the stylus on a Wacom tablet while attending to the system's display. Stylus tapping on a soft keyboard is a "direct" task: Participants manipulate the stylus on the soft keyboard while also visually attending to the soft keyboard. This, alone, is cause to suspect that the bandwidth coefficient used in Soukoreff and MacKenzie's model is conservative. Although no experiment measuring the bandwidth coefficient for stylus tapping on a soft keyboard has been published, we suspect such an experiment would yield a higher bandwidth. The effect would be to increase all the predictions in Figure 21.

Figure 24. Cubon keyboard (from Zhai, Hunter, & Smith, 2000).

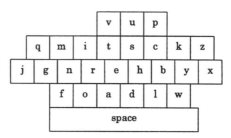

Figure 25. (a) OPTI I; (b) OPTI II soft keyboards. From "The design and evaluation of a high-performance soft keyboard," by I. S. MacKenzie and S. X. Zhang, 1999, *Proceedings of the CHI 99 Conference on Human Factors in Computing Systems.* Copyright 1999 by ACM. Reprinted with permission. From *A high performance soft keyboard for mobile systems,* by S. X. Zhang, 1998, Unpublished thesis, University of Guelph, Ontario, Canada. Copyright 1998 by S. X. Zhang. Reprinted with permission.

a)

q	f	u	m	c	k	z
space		o	t	h	space	
b	s	r	e	a	w	x
space		i	n	d	space	
j	p	v	g	l	y	

b)

q	k	c	g	v	j
space	s	i	n	d	space
w	t	h	e	a	m
space	u	o	r	l	space
z	b	f	y	p	x

In follow-up work, Zhang (1998) proposed a slight modification to OPTI I. The OPTI II appears in Figure 25b and yields an expert prediction of 42.37 wpm (see Figure 21).

Hunter et al. (2000) and Zhai et al. (2000) applied two physics-inspired techniques to the model of Soukoreff and MacKenzie and generated optimal keyboards. They used a mechanical simulation of a mesh of springs, where the springs were stretched between the characters of the alphabet and tensioned proportionally to digram probabilities in English. The technique is an application of a greedy algorithm to reduce the physical distance between more likely character pairs.[9] The result is a keyboard they call Hook's key-

---

9. The term *greedy algorithm* refers to a class of algorithms for solving minimization (or maximization) problems. Greedy algorithms try to find the minimum solution of a problem by always moving in the direction of steepest descent. However, greedy algorithms can become trapped in "local minimums." This is analogous to searching for the deepest point in a valley by always walking downhill—but becoming trapped in a hole in the side of the valley because a step upward (to a higher altitude) would be required to leave the hole.

board after Hook's law[10] (see Figure 26a). It yields a predicted expert entry rate of 41.15 wpm.

A second approach they took was to apply the Metropolis algorithm, which is theoretically more appealing because it employs a random-walk strategy instead of a greedy algorithm.[11] The Metropolis I and Metropolis II keyboards have higher predicted expert typing speeds and appear in Figure 26b and Figure 26c. In their first publication reporting preliminary results (Hunter et al., 2000), they presented a Metropolis keyboard, which we denote Metropolis I. In a later publication (Zhai et al., 2000), another Metropolis-derived keyboard was presented, which we call Metropolis II. The predicted expert entry rates for the Metropolis I and Metropolis II keyboards are 42.15 wpm and 42.94 wpm, respectively. Metropolis II has the distinction of yielding the fastest predictions of any soft keyboard tested (see Figure 21). Longitudinal evaluations of the Hook's, Metropolis I, or Metropolis II keyboards have not been undertaken, so the entry speeds attainable in practice are not known.

Lewis et al. (1999a, 1999b) also tried to optimize entry rates for a soft keyboard. They applied network analysis to character pair probabilities to determine the most strongly associated pairs. Then, using an ad hoc method to minimize distances for the strongly associated character pairs, they produced the keyboard arrangement in Figure 27. Lewis et al. performed a comparative user evaluation but they did not report their results; estimating from their published report (Lewis et al., 1999b, Figure 1) suggests they measured typing speeds of 25 wpm for the Qwerty control condition and 13 wpm for their keyboard design. They also reported that when asked, participants indicated a preference for the Qwerty layout. Our expert prediction for the Lewis keyboard is 34.65 wpm (see Figure 21).

The DotNote keyboard by Útilware (http://www.utilware.com) was designed to support single-handed text entry on the Palm as an alternative to the built-in Graffiti handwriting recognition, which requires two hands (one to hold the device, the other to manipulate the stylus). To support finger or

---

10. Hook's law states that the tensional force in a spring is proportional to its extension, that is, how much it has been stretched from its equilibrium length, or, $F = -kx$, where F is the force of the spring, x is the distance that the spring has been stretched, and k is the spring constant, which varies from spring to spring.

11. The Metropolis algorithm is a well-known approach to solving complex minimization (or maximization) problems, inspired by thermodynamics (Press, Teukolsky, Vetterling, & Flannery, 1992, p. 444). When a liquid is slowly cooled until it is solid, the resulting crystal is very orderly and has almost the minimum energy possible. Metropolis takes a function representing the energy of a system and applies simulated annealing solving the minimization problem by modeling the effect slow cooling has on the energy of the system. Metropolis does not suffer from the local minimum problem.

Figure 26. (a) Hook's keyboard; (b) Metropolis I keyboard; (c) Metropolis II keyboard. From "Physics-based graphical keyboard design," by M. Hunter, S. Zhai, and B. A. Smith, 2000, *Extended Abstracts of the CHI 2000 Conference on Human Factors in Computing Systems.* Copyright 2000 by ACM. Reprinted with permission. From "The Metropolis keyboard: An exploration of quantitative techniques for virtual keyboard design," by S. Zhai, M. Hunter, and B. A. Smith, 2000, *Proceedings of the ACM Conference on User Interface Software and Technology–UIST 2000.* Copyright 2000 by ACM. Reprinted with permission.

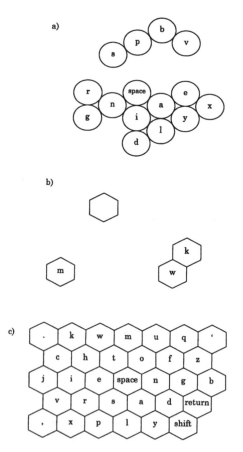

thumb typing, the DotNote keyboard fills most of the display with relatively large keys; however, this allows only half of the alphabet to appear at once. The most common letters appear on the default DotNote keyboard (Figure 28a) and a mode Shift key switches to the second keyboard arrangement, which contains the less common letters (Figure 28b). The arrangement of the keys in each soft keyboard is alphabetic. No published studies of the DotNote keyboard exist.

Figure 27. Lewis keyboard. From "Development of a digram-based typing key layout for single-finger/stylus input," by J. R. Lewis, M. J. LaLomia, and P. J. Kennedy, 1999, *Proceedings of the Human Factors and Ergonomics Society 43rd Annual Meeting-HFES 99.* Copyright 1999 by the Human Factors and Ergonomics Society. Reprinted with permission.

q	r	w	x	y	
l	u	a	o	f	
t	h	e	n	g	
v	d	i	s	p	
b	c	m	j	k	z
space					

Figure 28. DotNote soft keyboard: (a) the default key arrangement containing the more frequent letters; (b) the secondary key arrangement containing the less frequent letters.

a)

				←
A	C	D	E	⇧
G	H	I	L	123
M	N	O	P	A↔Q
R	S	T	U	Space

b)

B	F	J	K	
Q	V	W	X	123
Y	Z	.	,	A↔Q
"	–			

4.3. Predictive Input Techniques

One early predictive input technology is the Reactive Keyboard (Darragh, 1988; Darragh & Witten, 1991; Darragh, Witten, & James, 1990). The Reactive Keyboard monitors what a user enters and presents text predictions that the user can choose from using the mouse. The predictions are generated by finding the longest matching substrings in the previously entered text. The Reactive Keyboard adapts to users' input and hence is not limited to a static set of words or phrases. No experimental results of text entry speed or accuracy are reported for the Reactive Keyboard. Other related work is hereby cited (Jakobsson, 1986; Masui & Nakayama, 1994; Raita & Teuhola, 1987).

POBox (Masui, 1998, 1999) is predictive input technology that allows users to enter part of a word and then search for similar words by spelling, pronunciation, or shape (for pictograph-based languages). It is not limited to alphabetic languages. POBox uses a static database coupled with another primary input technique, such as a soft keyboard or telephone keypad. Search results appear on the screen as the user types. A tap or key press selects the desired word. When embedded in a mobile phone, text entry via the multikey method yields a list of search results that the user scrolls through using a wheel on the side of the device.

Lewis (1999) and Lewis et al. (1999) experimented with a predictive soft keyboard technology for extremely limited screen sizes. Their system presents the user with keys for the six to eight most likely characters, and an "other" key revealing the rest of the alphabet. Lewis (1999) reported text-entry speeds for a Qwerty soft keyboard, his predictive keyboard, and handwriting (as a control condition) at 14 wpm, 6 wpm, and 22 wpm, respectively. The speed Lewis reported for soft keyboard entry is approximately half that reported by others (e.g., MacKenzie et al., 1994). Lewis observed that the uncertain arrangement of keys on the predictive keyboard significantly hindered performance.

5. CONCLUSIONS AND FUTURE WORK

There are many text entry methods available to designers of mobile systems, and without a doubt more are forthcoming. However, deciding which is best for an application is difficult, in part, because of the lack of publications giving empirically measured text entry speeds and accuracies. This article has brought together many of the techniques in use or under investigation in this exciting area in mobile computing. The result is a snapshot of the current state of the art in mobile text entry.

Some important modeling techniques have been presented and elaborated. Movement and language are omnipresent in human–computer interaction. The Fitts' Digram Model shows how Fitts' law and a language corpus can work together in a priori analyses of design alternatives for stylus input on soft keyboards or single-finger input on small physical keyboards. However, further work is needed to refine this modeling technique—for example, in determining the correct coefficients for the Fitts' law model and in exploring and refining other aspects of the model, such as treatment of the space and punctuation characters or its sensitivity to changes in the language model.

In addition, we have examined many issues in methodology and evaluation and have identified factors, such as focus of attention, and whether one or two hands are used to manipulate the text entry device that impact user performance. Clearly, evaluation is critical, and it is by no means simple. A number of issues are particularly tricky, such as the measurement and treatment of errors and the types of tasks used in text entry studies. These and other topics are the subject of ongoing and future work.

NOTES

Acknowledgments. We thank Shumin Zhai, Barton Smith, and Michael Hunter for pointing out an error in the treatment of the Space character in our prediction model for soft keyboards. Their assistance—and persistence—in refining, improving, and extending our previous work is greatly appreciated.

Support. We gratefully acknowledge the support of the Natural Sciences and Engineering Research Council of Canada.

Authors' Present Addresses. Scott MacKenzie, Department of Computer Science, York University, Toronto, Ontario, Canada, M3J 1P3. E-mail: smackenzie@acm.org; William Soukoreff, Department of Computer Science, York University, Toronto, Ontario, Canada, M3J 1P3. E-mail: will@acm.org.

HCI Editorial Record. First manuscript received November 9, 2000. Accepted by Brad Myers. Final manuscript received June 1, 2001. — *Editor*

REFERENCES

Alsio, G., & Goldstein, M. (2000). Productivity prediction by extrapolation: Using workload memory as a predictor of target performance. *Behaviour & Information Technology, 19*(2), 87–96.

American Psychological Association. (1995). *Publication manual of the American Psychological Association* (4th ed.). Washington, DC: Author.

Bellman, T., & MacKenzie, I. S. (1998). A probabilistic character layout strategy for mobile text entry. *Proceedings of Graphics Interface 98.* Toronto: Canadian Information Processing Society.

Blickenstorfer, C. H. (1995, January). Graffiti: Wow!!!! *Pen Computing Magazine,* 30–31.

Bohan, M., Phipps, C. A., Chaparro, A., & Halcomb, C. G. (1999). A psychophysical comparison of two stylus-driven soft keyboards. *Proceedings of Graphics Interface 99.* Toronto: Canadian Information Processing Society.

Card, S. K., English, W. K., & Burr, B. J. (1978). Evaluation of mouse, rate-controlled isometric joystick, step keys, and text keys for text selection on a CRT. *Ergonomics, 21*, 601–613.

Card, S. K., Moran, T. P., & Newell, A. (1983). *The psychology of human–computer interaction*. Hillsdale, NJ: Lawrence Erlbaum Associates, Inc.

Chang, L., & MacKenzie, I. S. (1994). A comparison of two handwriting recognizers for pen-based computers. *Proceedings of CASCON 94*. Toronto, Canada: IBM.

Damerau, F. J. (1964). A technique for computer detection and correction of spelling errors. *Communications of the ACM, 7*(3), 171–176.

Darragh, J. J. (1988). *Adaptive predictive text generation and the Reactive Keyboard*. Unpublished thesis, University of Calgary, Canada.

Darragh, J. J., & Witten, I. H. (1991). Adaptive predictive text generation and the reactive keyboard. *Interacting With Computers, 3*(1), 27–50.

Darragh, J. J., Witten, I. H., & James, M. L. (1990). The reactive keyboard: A predictive typing aid. *Computer, 23*(11), 41–49.

Devoe, D. B. (1967). Alternatives to handprinting in the manual entry of data. *IEEE Transactions on Human Factors in Engineering, HFE-8*, 21–32.

Dix, A., Finlay, J., Abowd, G., & Beale, R. (1998). *Human–computer interaction* (2nd ed.). London: Prentice Hall.

Enns, N., & MacKenzie, I. S. (1998). Touch-based remote control devices. *Extended Abstracts of the CHI 98 Conference on Human Factors in Computing Systems*. New York: ACM.

Fitts, P. M. (1954). The information capacity of the human motor system in controlling the amplitude of movement. *Journal of Experimental Psychology, 47*, 381–391.

Frankish, C., Hull, R., & Morgan, P. (1995). Recognition accuracy and user acceptance of pen interfaces. *Proceedings of the CHI 95 Conference on Human Factors in Computing Systems*. New York: ACM.

Fukumoto, M., & Suenaga, Y. (1994). FingeRing: A full-time wearable interface. *Proceedings of CHI 94 Conference on Human Factors in Computing Systems*. New York: ACM.

Gentner, D. R., Grudin, J. T., Larochelle, S., Norman, D. A., & Rumelhart, D. E. (1983). A glossary of terms including a classification of typing errors. In W. E. Cooper (Ed.), *Cognitive aspects of skilled typing* (pp. 39–44). New York: Springer-Verlag.

Gibbs, M. (1993). Handwriting recognition: A comprehensive comparison. *Pen*, (March/April), 31–35.

Goldberg, D., & Richardson, C. (1993). Touch-typing with a stylus. *Proceedings of the INTERCHI 93 Conference on Human Factors in Computing Systems*. New York: ACM.

Goldstein, M., Book, R., Alsio, G., & Tessa, S. (1999). Non-keyboard Qwerty touch typing: A portable input interface for the mobile user. *Proceedings of the CHI 99 Conference on Human Factors in Computing Systems*. New York: ACM.

Gopher, D., & Raij, D. (1988). Typing with a two-hand chord keyboard: Will the Qwerty become obsolete? *IEEE Transactions of Systems, Man, and Cybernetics, 18*, 601–609.

Grinter, R. E., & Eldridge, M. A. (2001). y do tngrs luv 2 txt msg? *Proceedings of the ECSCW 2001 European Conference on Computer Supported Collaborative Work*. Dordrecht, The Netherlands: Kluwer Academic.

Grover, D. L., King, M. T., & Kuschler, C. A. (1998). *Reduced keyboard disambiguating computer.* Seattle, WA: Tegic Communications.

Hancock, P. A., & Newell, K. M. (1985). The movement speed–accuracy relationship in space-time. In H. Heuer, U. Kleinbeck, & K.-H. Schmidt (Eds.), *Motor behavior: Programming, control and acquisition* (pp. 153–188). New York: Springer-Verlag.

Hick, W. E. (1952). On the rate of gain of information. *Quarterly Journal of Experimental Psychology, 4,* 11–36.

Hjelm, J., Tan, C. L., Fabry, L., Fanchon, T., & Reichert, F. (1996). Building the UMTS user interface. *Advanced Communications Technologies and Services (ACTS) Mobile Communications Summit* (Vol. 2, pp. 687–692). Brussels, Belgium: Acts European Commission.

Hunter, M., Zhai, S., & Smith, B. A. (2000). Physics-based graphical keyboard design. *Extended Abstracts of the CHI 2000 Conference on Human Factors in Computing Systems.* New York: ACM.

Hyman, R. (1953). Stimulus information as a determinent of reaction time. *Journal of Experimental Psychology, 45,* 188–196.

Isokoski, P. (1999). *A minimal device-independent text input method.* Unpublished thesis, University of Tampere, Finland.

Jakobsson, M. (1986). Autocompletion in full text transcription entry: A method for humanized input. *Proceedings of the CHI 86 Conference on Human Factors in Computing Systems.* New York: ACM.

Kay, A., & Goldberg, A. (1977, March). Personal dynamic media. *Computer,* 31–41.

LaLomia, M. J. (1994). User acceptance of handwritten recognition accuracy. *Companion Proceedings of the CHI 94 Conference on Human Factors in Computing Systems.* New York: ACM.

Levenshtein, V. I. (1966). Binary codes capable of correcting deletions, insertions and reversals. *Soviet Physics–Doklady, 10,* 707–710.

Lewis, J. R. (1999). Input rates and user preference for three small-screen input methods: Standard keyboard, predictive keyboard, and handwriting. *Proceedings of the Human Factors and Ergonomics Society 43rd Annual Meeting. Santa Monica, CA: Human Factors and Ergonomics Society.*

Lewis, J. R., Allard, D. J., & Hudson, H. D. (1999). Predictive keyboard design study: Effects of word populations, number of displayed letters, and number of transitional probability tables. *Proceedings of the Human Factors and Ergonomics Society 43rd Annual Meeting.* Santa Monica, CA: Human Factors and Ergonomics Society.

Lewis, J. R., LaLomia, M. J., & Kennedy, P. J. (1999a). Development of a digram-based typing key layout for single-finger/stylus input. *Proceedings of the Human Factors and Ergonomics Society 43rd Annual Meeting.* Santa Monica, CA: Human Factors and Ergonomics Society.

Lewis, J. R., LaLomia, M. J., & Kennedy, P. J. (1999b). Evaluation of typing key layouts for stylus input. *Proceedings of the Human Factors and Ergonomics Society 43rd Annual Meeting. Santa Monica, CA: Human Factors and Ergonomics Society.*

MacKenzie, I. S. (1992). Fitts' law as a research and design tool in human–computer interaction. *Human–Computer Interaction, 7,* 91–139.

MacKenzie, I. S., & Chang, L. (1999). A performance comparison of two handwriting recognizers. *Interacting with Computers, 11,* 283–297.

MacKenzie, I. S., Nonnecke, R. B., McQueen, J. C., Riddersma, S., & Meltz, M. (1994). A comparison of three methods of character entry on pen-based computers. *Proceedings of the Human Factors Society 38th Annual Meeting.* Santa Monica, CA: Human Factors and Ergonomics Society.

MacKenzie, I. S., Nonnecke, R. B., Riddersma, S., McQueen, C., & Meltz, M. (1994). Alphanumeric entry on pen-based computers. *International Journal of Human-Computer Studies, 41,* 775–792.

MacKenzie, I. S., Sellen, A., & Buxton, W. (1991). A comparison of input devices in elemental pointing and dragging tasks. *Proceedings of the CHI 91 Conference on Human Factors in Computing Systems.* New York: ACM.

MacKenzie, I. S., & Zhang, S. X. (1997). The immediate usability of Graffiti. *Proceedings of Graphics Interface 97.* Toronto: Canadian Information Processing Society.

MacKenzie, I. S., & Zhang, S. X. (1999). The design and evaluation of a high-performance soft keyboard. *Proceedings of the CHI 99 Conference on Human Factors in Computing Systems.* New York: ACM.

MacKenzie, I. S., & Zhang, S. X. (2000). *An investigation of the novice experience with soft keyboards.* Manuscript submitted for publication.

MacKenzie, I. S., Zhang, S. X., & Soukoreff, R. W. (1999). Text entry using soft keyboards. *Behaviour & Information Technology, 18,* 235–244.

Mankoff, J., & Abowd, G. A. (1998). Cirrin: A word-level unistroke keyboard for pen input. *Proceedings of the UIST 98 Symposium on User Interface Software and Technology.* New York: ACM.

Martin, D. W. (1996). *Doing psychology experiments* (4th ed.). Pacific Grove, CA: Brooks/Cole.

Masui, T. (1998). An efficient text input method for pen-based computers. *Proceedings of the CHI 98 Conference on Human Factors in Computing Systems.* New York: ACM.

Masui, T. (1999). POBox: An efficient text input method for handheld and ubiquitous computers. *Proceedings of the HUC 99 International Symposium on Handheld and Ubiquitous Computing.* Berline, Germany: Springer-Velag.

Masui, T., & Nakayama, K. (1994). Repeat and predict: Two keys to efficient text editing. *Proceedings of the CHI 94 Conference on Human Factors in Computing Systems.* New York: ACM.

Matias, E., MacKenzie, I. S., & Buxton, W. (1993). Half-Qwerty: A one-handed keyboard facilitating skill transfer from Qwerty. *Proceedings of the INTERCHI 93 Conference on Human Factors in Computing Systems.* New York: ACM.

Matias, E., MacKenzie, I. S., & Buxton, W. (1994). Half-Qwerty: Typing with one hand using your two-handed skills. *Companion Proceedings of the CHI 94 Conference on Human Factors in Computing Systems.* New York: ACM.

Matias, E., MacKenzie, I. S., & Buxton, W. (1996a). One-handed touch typing on a Qwerty keyboard. *Human-Computer Interaction, 11,* 1–27.

Matias, E., MacKenzie, I. S., & Buxton, W. (1996b). A wearable computer for use in microgravity space and other non-desktop environments. *Companion Proceedings of the CHI 96 Conference on Human Factors in Computing Systems.* New York: ACM.

Mayzner, M. S., & Tresselt, M. E. (1965). Table of single-letter and digram frequency counts for various word-length and letter-position combinations. *Psychonomic Monograph Supplements, 1*(2), 13–32.

McMulkin, M. (1992). Description and prediction of long-term learning of a keyboarding task. *Proceedings of the Human Factors Society 36th Annual Meeting.* Santa Monica, CA: Human Factors Society.

Pachella, R. G., & Pew, R. W. (1968). Speed–accuracy tradeoff in reaction-time: Effect of discrete criterion times. *Journal of Experimental Psychology, 76*(1), 19–24.

Perlin, K. (1998). Quikwriting: Continuous stylus-based text entry. *Proceedings of the UIST 98 Symposium on User Interface Software and Technology.* New York: ACM.

Pew, R. W. (1969). The speed–accuracy operation characteristic. In W. G. Koster (Ed.), *Attention and performance II* (pp. 16–26). Amsterdam: North-Holland.

Poon, A., Weber, K., & Cass, T. (1995). Scribbler: A tool for searching digital ink. *Companion Proceedings of the CHI 95 Conference on Human Factors in Computing Systems.* New York: ACM.

Press, W. H., Teukolsky, S. A., Vetterling, W. T., & Flannery, B. P. (1992). *Numerical recipes in C: The art of scientific computing* (2nd ed.). Cambridge, England: Cambridge University Press.

Raita, T., & Teuhola, J. (1987). Predictive text compression by hashing. *Proceedings of the Tenth Annual International ACM SIGIR Conference on Research and Development in Information Retrieval.* New York: ACM.

Rau, H., & Skiena, S. (1994). Dialing for documents: An experiment in information theory. *Proceedings of the UIST 94 Symposium on User Interface Software and Technology.* New York: ACM.

Sacher, H. (1998, September/October). Interactions in Chinese: Designing interfaces for Asian languages. *Interactions,* 28–38.

Santos, P. J., Baltzer, A. J., Badre, A. N., Henneman, R. L., & Miller, M. S. (1992). On handwriting recognition system performance: Some experimental results. *Proceedings of the Human Factors Society 36th Annual Meeting.* Santa Monica, CA: Human Factors and Ergonomics Society.

Shannon, C. E. (1951). Prediction and entropy of printed English. *Bell System Technical Journal, 30,* 51–64.

Shneiderman, B. (2000, September). The limits of speech recognition. *Communications of the ACM,* 63–65.

Silfverberg, M., MacKenzie, I. S., & Korhonen, P. (2000). Predicting text entry speed on mobile phones. *Proceedings of the CHI 2000 Conference on Human Factors in Computing Systems.* New York: ACM.

Soukoreff, R. W., & MacKenzie, I. S. (1995). Theoretical upper and lower bounds on typing speeds using a stylus and soft keyboard. *Behaviour & Information Technology, 14,* 370–379.

Soukoreff, R. W., & MacKenzie, I. S. (2001). Measuring errors in text entry tasks: An application of the Levenshtein string distance statistic. *Companion Proceedings of the CHI 2001 Conference on Human Factors in Computing Systems.* New York: ACM.

Sugimoto, M., & Takahashi, K. (1996). SHK: Single hand key card for mobile devices. In *Companion Proceedings of the CHI 96 Conference on Human Factors in Computing Systems.* New York: ACM.

Suhm, B., Myers, B., & Waibel, A. (1999). Model-based and empirical evaluation of multimodal interactive error correction. *Proceedings of the CHI 99 Conference on Human Factors in Computing Systems.* New York: ACM.

Swensson, R. G. (1972). The elusive tradeoff: Speed vs. accuracy in visual discrimination tasks. *Perception and Psychophysics, 12*(1A), 16–32.

Tappert, C. C., Suen, C. Y., & Wakahara, T. (1990). The state of the art in on-line handwriting recognition. *IEEE Transactions on Pattern Analysis and Machine Intelligence, 12*, 787–808.

Trudeau, G. B. (1996). *Flashbacks: Twenty-five years of Doonesbury.* Kansas City, MO: Andrews McMeel.

Underwood, B. J., & Schulz, R. W. (1960). *Meaningfulness and verbal learning.* Philadelphia: Lippincott.

Van Cott, H. P., & Kinkade, R. G. (Eds.). (1972). *Human engineering guide to equipment design.* Washington, DC: U.S. Government Printing Office.

Venolia, D., & Neiberg, F. (1994). T-Cube: A fast, self-disclosing pen-based alphabet. *Proceedings of CHI 94 Conference on Human Factors in Computing Systems.* New York: ACM.

Walker, N., & Catrambone, R. (1993). Aggregation bias and the use of regression in evaluation models of human performance. *Human Factors, 35*, 397–411.

Ward, D. J., Blackwell, A. F., & MacKay, D. J. C. (2000). Dasher: A data entry interface using continuous gestures and language models. *Proceedings of the UIST 2000 Symposium on User Interface and Software Technology.* New York: ACM.

Ward, D. J., Blackwell, A. F., & MacKay, D. J. C. (2002). Dasher: A gesture-driven data entry interface for mobile computing. *Human–Computer Interaction, 17*, 199–228.

Wickelgren, W. A. (1977). Speed–accuracy tradeoff and information processing dynamics. *Acta Psychologica, 41*, 67–85.

Wolf, C. G., Glasser, A. R., & Fujisaki, T. (1991). An evaluation of recognition accuracy for discrete and run-on writing. *Proceedings of the Human Factors Society 35th Annual Meeting.* Santa Monica, CA: Human Factors and Ergonomics Society.

Zhai, S., Hunter, M., & Smith, B. A. (2000). The Metropolis keyboard: An exploration of quantitative techniques for virtual keyboard design. *Proceedings of the UIST 2000 Symposium on User Interface Software and Technology.* New York: ACM.

Zhang, S. X. (1998). *A high performance soft keyboard for mobile systems.* Unpublished thesis, University of Guelph, Ontario, Canada.

HUMAN-COMPUTER INTERACTION, 2002, Volume 17, pp. 199–228
Copyright © 2002, Lawrence Erlbaum Associates, Inc.

Dasher: A Gesture-Driven Data Entry Interface for Mobile Computing

David J. Ward, Alan F. Blackwell, and David J. C. MacKay
University of Cambridge

ABSTRACT

Existing devices for communicating information to computers are bulky, slow, or unreliable. Dasher is an interface incorporating language modeling and driven by continuous two-dimensional gestures (e.g., a mouse, a stylus, or eye-tracker). Tests have shown that, after 1 hr of practice, novice users reach a writing speed of about 20 words per minute (wpm) while taking dictation. Experienced users achieve writing speeds of about 34 wpm, compared with typical 10-finger keyboard typing of 40 to 60 wpm.

Although the interface is slower than a conventional keyboard, it is simple to use and could be used on personal data assistants and by motion-impaired computer users. Dasher can readily be used to enter text from any alphabet.

David Ward is a physicist with interests in human–computer interaction, language modeling, and neural networks; he is a graduate student in the Department of Physics at the University of Cambridge. Alan Blackwell is a psychologist and engineer with interests in human–computer interaction and software product design; he is a lecturer in the Computer Laboratory at the University of Cambridge. David MacKay is a physicist with interests in machine learning and information theory; he is a reader in the Department of Physics at the University of Cambridge.

CONTENTS

1. THE INFORMATION CONTENT OF TEXT AND OF HAND MOVEMENTS

Existing devices for communicating information to computers are bulky, slow to use, or unreliable. Dasher is a new interface incorporating language modeling and driven by continuous two-dimensional gestures (e.g., a mouse, touch screen, or eye-tracker). Dasher is easy to learn, attains moderate writing speeds, and is small enough to be implemented on personal data assistants (PDAs).

Conventional keyboards require the user to make one or two gestures per character (one for lowercase and two for uppercase characters). Each gesture is a selection of 1 from 80 keys, so the keyboard has the capacity to read about $\log_2(80) = 6.3$ bits per gesture. However, the entropy of English text is roughly 1 bit per character (bpc; Shannon, 1948), so existing keyboards are inefficient by a factor of 6. This inefficiency manifests itself in the fact that when typing, people make errors that are clearly not English.

The keyboard is inefficient for two reasons. First, text is usually highly redundant, yet users are forced to type most or all of the characters in a document. Second, keyboards register only contact events, whereas humans are capable of fine motor movements. Human motor capabilities dictate an upper limit for finger tapping frequency and hence the information rate achievable from successive key presses, no matter how small the distance between keys. Continuous pointing gestures have the potential to convey more information through high-resolution measurement of position.

According to MacKenzie's (1992a) analysis of Fitts' experiments measuring one-dimensional pointing speed and accuracy, the rate at which information can be conveyed by one-dimensional pointing is about 8.2 bits per second (bps). So using just one finger in a one-dimensional pointing environment, it should be possible to write English at a rate of 8.2 characters per second. For this to be achieved, we must connect the finger to a good probabilistic model of the English language. This character rate corresponds to about 100 words per minute (wpm). How close we can get to this figure depends on the quality of our model of English and on whether we can map finger gestures to text in a way that people can easily learn.

2. RELATION TO PREVIOUS WORK

Personal organizers with touch-sensitive screens are small and convenient to carry, but at present, data are usually entered using miniature keyboards or a slow handwriting system. Speech recognition systems are almost as fast as conventional keyboards (Jecker, 1999) but are not always socially acceptable in a crowded office, and users may also suffer from voice fatigue.

Gesture-based text entry systems involve two significant trade-offs between potential efficiency and training time and between device size and character-set size. Research on keyboards for conventional typing has addressed both efficiency versus training (Norman & Fisher, 1982) and the effect of keyboard size (Sears, Revis, Swatski, Crittenden, & Shneiderman, 1993). In conventional keyboards, efficiency depends on relative key size (according to Fitts' law; MacKenzie, 1992b) and on the mapping between movement over the keyboard configuration and English letter sequences. Learning speed appears to be a function of similarity to the QWERTY layout (Gordon, Henry, & Massengill, 1975), which has almost ubiquitous familiarity, even in nontypists (Norman & Fisher, 1982).

The difficulty of resolving these trade-offs has led to predominance of the Qwerty keyboard, even in the face of mechanical improvements such as the Dvorak keyboard. However, in the case of handheld devices no single text entry system is dominant.

2.1. Previous Solutions

Approaches to text entry in handheld devices can be grouped into three broad categories: miniature and rearranged keyboards, gestural alphabets, and dynamic selection techniques.

Miniature and Rearranged Keyboards

The challenge in this approach is to present as much of the alphabet as possible while retaining sufficiently large keys. The number of keys is often reduced, and some selection mechanism is used to toggle between alternate sets. The Half-Qwerty device uses half of the Qwerty keyboard, and the Space bar toggles between the two halves. Evaluation showed typing speeds of 34.7 wpm after 10 sessions, each lasting 50 min (Matias, MacKenzie, & Buxton, 1993). The FITALY keyboard arranges keys in a compact square optimized for minimal hand travel with one-finger typing (MacKenzie, Zhang, & Soukoreff, 1999).

The T9® text input system uses the conventional mapping of the 26 alphabetic characters onto a 12-key telephone keypad, with a dictionary to help disambiguate alternative readings (Silfverberg, MacKenzie, & Korhonen, 2000). Both FITALY and T9 are available as alternative text entry systems for the Palm™ Pilot. T9 is also increasingly well known as a predictive text entry system used in mobile phones. An experiment with 20 participants compared novice and expert mobile phone text users (James & Reischel, 2001) using two different styles of text: chat messages and newspaper extracts. For chat messages, experts achieved 26 wpm with T9 versus 11 wpm for multitap. Novices achieved 11 wpm with T9 versus 10 wpm with multitap.

A mathematically optimized keyboard has been proposed in which keys are arranged on a hexagonal grid (Hunter, Zhai, & Smith, 2000). Predicted text entry speed is 42 to 44 wpm. Recent evaluation (Zhai & Smith, 2001) reports writing speeds of 8.9 to 9.7 wpm.

Some evaluation has been done to measure the effect of using standard Qwerty keyboard layouts reduced in size on a one-finger touch screen (Sears et al., 1993). Typing speed for experts after 30 min was 32.5 wpm with a large keyboard (246 mm wide), and dropped to 21.1 wpm on a very small one (68 mm wide).

It is also possible to reduce keyboard size with "chorded" schemes having fewer keys but in which multiple keys are pressed at one time. Performance of the ternary chorded keyboard (TCK; Kroemer, 1992) was about 16 wpm after 600 min of training.

Gestural Alphabets

Most handheld pen devices are supplied with software to recognize hand-written characters, possibly using a special alphabet. The gesture sets range from those that are designed to be very efficient to those that emphasize ease of learning. Early devices attempted, with limited success, to recognize users' natural handwriting and hence required no learning on the part of the user (as in the case of the Apple® Newton®). Current systems—Graffiti® (MacKenzie & Zhang, 1997) for the Palm Pilot and Jot for Windows® CE (also available on the Pilot)—improve efficiency while maintaining some resemblance to alphabetic shapes to assist learning. Researchers have also proposed more efficient alphabets such as Unistrokes, which maps simple gestures to common characters regardless of mnemonic similarity (Goldberg & Richardson, 1993). The inventors of Unistrokes estimate a rate of 40.8 wpm, without any empirical testing.

Dynamic Selection

Dynamic selection techniques share characteristics of miniature keyboards and gestural alphabets. The user moves the pen in the direction of a selection region, which either selects a character or reveals a further set of alternatives. Examples include T-Cube (Venolia & Neiberg, 1994) and Quikwriting (Perlin, 1998). Quikwriting is commercially available for the Palm Pilot.

An unusual alternative approach has been to start with a standard Qwerty keyboard but then rearrange the keys after each keystroke to place the most probable next letters near the center. Results with the fluctuating optimal character layout keyboard were around 10 wpm after 10 sessions of 15 min (Bellman & MacKenzie, 1998).

These dynamic selection devices are subject to a trade-off between efficiency and ease of learning. The arrangement of nested regions or sets of alternatives must favor the selection of common characters, while still being learnable. Dasher offers an alternative dynamic selection technique that is easier to learn but improves efficiency by dynamically setting the size of the selection targets according to a language model. The system can be used with arbitrary alphabets; special characters and capitalization can be included without modification.

2.2. Previous Work on Dasher

We first described Dasher version 1.3 in Ward, Blackwell, and MacKay (2000). For completeness, this article includes the results from that article. In addition, this article includes (a) application to mobile computing, (b) details of the dynamics of Dasher, (c) modifications included in Dasher 1.5, (d) alternative character sets, and (e) theoretical analysis of maximum information rate. The current version of Dasher is 1.6.

3. DASHER

3.1. Entering Text

We first describe how Dasher is used to enter the word *the*. Figure 1(a) shows the initial configuration, with an alphabet of 27 characters displayed alphabetically in a column. There are 26 lowercase letters, and the symbol "_" represents a space. The user writes the first letter by making a gesture toward the letter's rectangle. The trails show the user moving the mouse toward the letter *t*.

In a continuous motion, the point of view zooms toward this letter (Figure 1b). As the rectangles get larger, possible extensions of the written string appear within the rectangle that we are moving toward. So if we are moving into the *t*, rectangles corresponding to *ta, tb, ..., th, ..., tz* appear in a vertical line like the first line. The heights of the rectangles correspond to the probabilities of these strings, according to the language model. In English, *ta* is quite probable, *tb* is less so, and *th* is very probable. So it is easy to gesture our point of view into *th* (Figure 1c) and from there into *the* (Figure 1d).

Dasher has a Web site, http://www.inference.phy.cam.ac.uk, which has a noninteractive demo. Dasher can be downloaded for Microsoft® Windows and UNIX® platforms.

Figure 1. (a) Initial configuration; (b), (c), and (d) are three successive screen shots from Dasher showing the string *the* being entered. The symbol "_" represents a space.

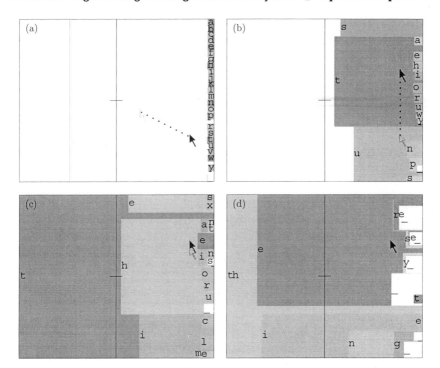

3.2. How the Probabilistic Model Determines the Screen Layout

In a given context, we display the alphabet of possible continuations as a column of characters as shown in Figure 2. The division of the right-hand vertical is analogous to arithmetic coding (Bell, Cleary, & Witten, 1990). Let our alphabet be $A_X = \{a_1, a_2, \ldots, a_I\}$. We divide the real line $[0,1)$ into I intervals of lengths equal to the probabilities $P(x_i = a_i)$.

We subdivide the interval a_i into intervals denoted $a_i a_1, a_i a_2, a_i a_3, \ldots, a_i a_I$, such that the length of the interval $a_i a_j$ is

$$P(x_1 = a_i, x_2 = a_j) = P(x_1 = a_i)\, P(x_2 = a_j | x_1 = a_i) \qquad (1)$$

The language model described in Section 5 assigns the probabilities. The interval $[0,1)$ can be divided into a sequence of intervals corresponding to all possible finite length strings $x_1 x_2 \ldots x_n$, such that the length of an interval is equal to the probability of the string given our model.

Figure 2. Arithmetic coding.

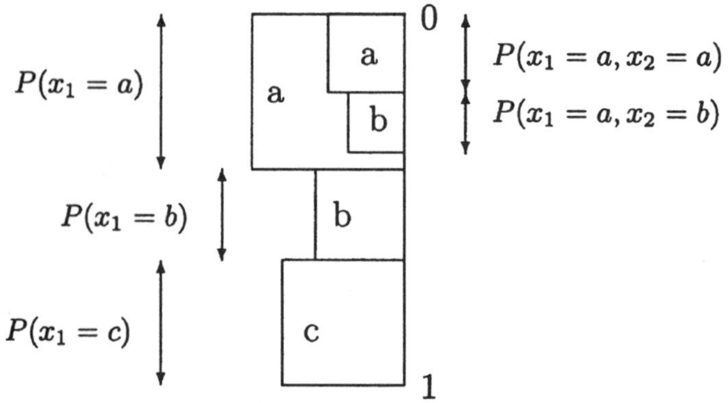

This sequence of intervals corresponds to the alphabetically ordered sequence of books in Borges's (2000) *Library of Babel*, with the size of each book being proportional to the probability of its contents under the language model. The user writes by gliding into the library and selecting the desired book.

In Figure 2, the rectangles are shown as squares, but we can use rectangles with any aspect ratio.

3.3. Dynamics of the Interface

The Steps Parameter, S

The pointing device controls a continuous zooming-in of the users' point of view. The interface zooms in so that the place under the pointer passes through the crosshair S frame updates later (Figure 3).

The left–right coordinate controls the rate of zooming-in, and the vertical coordinate determines the point on the right-hand vertical that is being zoomed into. The intuitive idea is that the user should point where they want to go.

Dynamical Equations

We use two coordinate systems to describe the dynamics of the interface. The first coordinate system specifies the visible interval on the real line (introduced in Section 3.2). The second are the screen coordinates.

In the initial configuration, the visible interval is [0,1). The visible interval after i screen updates can be specified by the center of the interval, C_i, and the height, H_i (Figure 4). In the initial configuration, the following equations apply:

Figure 3. Dynamics of the interface, assuming the user holds the pointer at a constant screen location. After *S* frame updates, the point of view has zoomed in so that the point under the mouse is now under the crosshair.

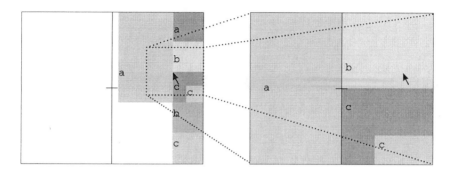

Figure 4. Screen update. The left diagram shows a view of the whole interval. The point of view is specified by a visible interval on the real line. The interval has a center, C_i, and a height, H_i. The left figure shows three subsequent intervals and on the right, the point of view.

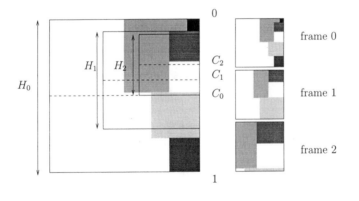

$$C_0 = .5 \tag{2}$$
$$H_0 = 1 \tag{3}$$

Figure 5 shows the coordinate system for the pointer (m_x, m_y). The position of the crosshair is $(k_x, .5)$. The following dynamical equations give the next interval, (C_{i+1}, H_{i+1}), in terms of the current interval, (C_i, H_i):

$$r = \frac{m_x}{k_x}, \tag{4}$$

$$r' = r^{1/S}, \tag{5}$$

Figure 5. Pointer coordinates. The height of the display is one screen unit.

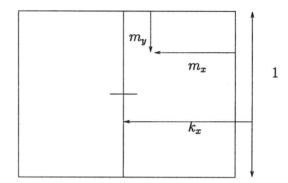

$$H_{i+1} = r'H_i, \tag{6}$$

$$C_{i+1} = C_i + H_i(0.5 - m_y)\frac{1-r'}{1-r}. \tag{7}$$

Setting the Steps Parameter, S

The steps parameter determines how responsive the interface is; the smaller it is, the faster the point of view can be changed. Inexperienced users usually require a relatively large value of S, say 70. A high value of S makes the system relatively unresponsive to large or inaccurate movements. As users improve, S can be decreased, enabling higher writing speeds.

However, when measuring their writing speed, we decided not to allow users to set S, otherwise their performance would be subject to their ability to choose this parameter as well as their skill in using the interface.

The speed of Dasher depends on the frame rate and CPU power so the appropriate value of S is system dependent. We designed an online algorithm to set S to a suitable value. Let x be the normalized distance of the pointer from the right-hand side of the screen,

$$x = \frac{m_x}{k_x}. \tag{8}$$

This normalization ensures that x is scale invariant.

Figure 6 qualitatively shows a typical user's response to different values of S. At point A on the graph, S is relatively small. The interface is too responsive, and the user keeps the pointer close to the crosshair, $x = 1$. At B, S is relatively large. The user is able to hold the pointer near the right-hand vertical, $x = 0$.

Figure 6. **Mean horizontal pointer position, $\bar{x}$. The graph qualitatively shows a typical user's response to different values of S.**

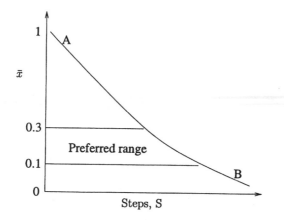

Experience shows that good writing speeds are obtained in the range $.1 < \bar{x} < .3$. Therefore, we used the following update rule for S;

$$\text{if } \bar{x} > .3, \; S \rightarrow S + 1 \tag{9}$$
$$\text{if } \bar{x} < .1, \; S \rightarrow S - 1 \tag{10}$$

We alter the timescale of adaptation by adjusting the size of the sample, from which $\bar{x}$ is computed. In practice, the timescale was around 15 sec.

3.4. Correcting Mistakes

Inexperienced users of Dasher often inadvertently follow a path other than that corresponding to the desired text. This generally happens when the desired string contains a high-probability string followed by a low-probability string. The user may react too late, inadvertently continuing forward motion to accept a high-probability string in the neighborhood of the desired text. After entering a high-probability neighbor, the desired text may not even be visible.

There are a number of ways in which we could handle corrections. Currently, the dynamics allow the user to "back up" by pointing to the left of the crosshair. The higher the leftward offset from the crosshair, the faster the zooming-out.

In Figure 7, the user has written the string *might* instead of *mean*. To alter this string, the user points to the left side of the display. Then the user points at the desired string, *mean*. An alternative mechanism, not yet tested, is to use an extra character to indicate that a misspelling has occurred.

Figure 7. Correcting mistakes. The user has written *might* instead of *mean*. In (a), the user points at *m* to go back. In (b), the user points at *mean*. The open arrowhead indicated the start of the motion; the closed arrowhead indicated the end.

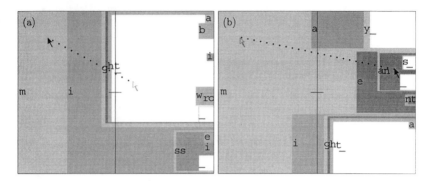

3.5. Benefits of Continuous Gestures and Language Modeling

Dasher is driven by continuous gestures, so inaccurate gestures can be compensated for by later gestures, the way a driver keeps a car on the road. When using a conventional keyboard we select one character per gesture, but in Dasher some gestures select more than one character, and therefore Dasher has the potential to convey information at higher rates than a keyboard. Figure 8 shows completions of the string *object*. Grammatically correct continuations can easily be selected with a single gesture.

The language model makes spelling mistakes less likely.

3.6. Horizontally Modified Display

When testing Dasher version 1.0, many users found that they were not able to see an adequate history and that the rate of zooming-out was too slow.

In Figure 9 the user writes *laboratory*, and the letters *la* pass out of the display before finishing the word. In version 1.5, we put the horizontal coordinates of the squares through a nonlinear mapping before rendering them on the screen. This mapping is linear on the right-hand side but logarithmic on the left-hand side. The effect can be seen by comparing Figures 9(a) and 9(b), which display the same interval on the real line. This display makes it easier for users to go back and correct mistakes.

If the user points to the *l* of *laboratory* using the modified display, the *l* will pass through the crosshair in *S* frame updates. With a square aspect ratio, it will take more frame updates to reach the *l* when pointing at the same screen coordinates. Therefore, the modified display gives a higher rate of zooming-out.

Figure 8. A single gesture can select several characters. Continuations of the string *object* include *objection*, *objects*, *object_and*, and *object_of.*

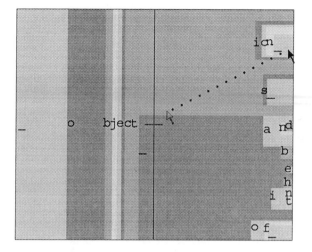

Figure 9. The user writes *laboratory*. (a) Square aspect ratio; (b) Modified display obtained by a nonlinear mapping of the horizontal coordinates.

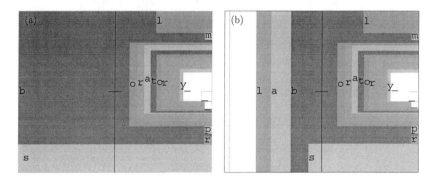

3.7. Vertically Modified Display

We also put the vertical coordinates through a nonlinear transformation, as shown in Figure 10. This has advantages of increasing the maximum speed of vertical scrolling and giving users more time to react before a desired letter disappears out of view.

4. ALTERNATIVE CHARACTER SETS

It is easy to incorporate any character set into Dasher. We give two examples.

Figure 10. Figure to demonstrate vertically modified display. (a) Square aspect ratio; (b) Modified display obtained by a nonlinear mapping of the vertical coordinates.

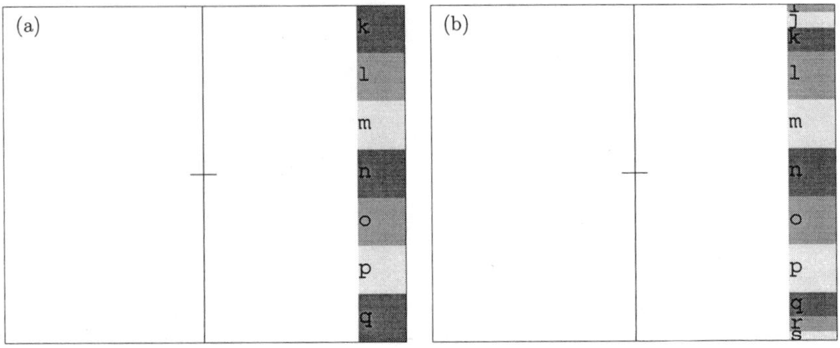

4.1. Capitalization

We can extend the character set introduced in Section 3.1 to 53 symbols by adding capital letters (Figure 11). In Dasher version 1.5, the lower- and upper-case letter are interleaved. When an expert user took dictation, they wrote at 42 wpm with lowercase and 41 wpm with both upper- and lowercase. It appears that the performance of Dasher is not significantly altered, even though the size of the character set has doubled. Notice how the language model predicts capitals where they are probable (after "Mr_") and lowercase letters after "Mr_K," "Mr_J," and so forth.

4.2. Japanese

Japanese is written in three different scripts.

- Kanji are Chinese characters or ideograms. Each character represents a word.
- Hiragana is a system in which each symbol represents a spoken syllable.
- Katakana is a phonetic alphabet, similar to Hiragana, but used for writing words of non-Japanese origin.

There are thousands of Kanji symbols, so for our first prototype, we decided to use the Hiragana character set. In the future we plan to make a combined Hiragana–Kanji system in which Kanji characters are selected by their relative probability as the completion of a phonetic Hiragana spelling. Hiragana consists of 83 symbols, as shown in Figure 12. The symbols are listed in their conventional order for computers.

Figure 11. Dasher with capital letters.

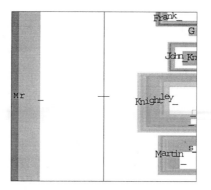

Figure 12. Shift-JIS character set for Hiragana.

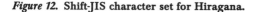

	0	1	2	3	4	5	6	7	8	9	A	B	C	D	E	F
0x8290															あ	
0x82A0	あ	ぃ	い	ぅ	う	ぇ	え	ぉ	お	か	が	き	ぎ	く	ぐ	け
0x82B0	げ	こ	ご	さ	ざ	し	じ	す	ず	せ	ぜ	そ	ぞ	た	だ	ち
0x82C0	ぢ	っ	つ	づ	て	で	と	ど	な	に	ぬ	ね	の	は	ば	ぱ
0x82D0	ひ	び	ぴ	ふ	ぶ	ぷ	へ	べ	ぺ	ほ	ぼ	ぽ	ま	み	む	め
0x82E0	も	ゃ	や	ゅ	ゆ	ょ	よ	ら	り	る	れ	ろ	ゎ	わ	ゐ	ゑ
0x82F0	を	ん														

Although Hiragana keyboards are available, many Japanese users type on conventional Qwerty keyboards. Each Hiragana symbol has a Romanization (e.g., *ka, ki, ku*). Typically, two key presses are required to enter one Hiragana symbol.

The Hiragana version, "Jdasher," shown in Figure 13, is incorporated into Dasher versions 1.5 onward. One ordering of characters is identical to that shown in Figure 12. The character set can be split into 10 classes, based on sound. For example, the characters *ka, ki, ku, ke, ko* and *sa, shi, su, se, so* form two different classes. In Dasher, we gave each class a fixed color to aid the user's search. In an alternative character set, we separated the diacritical marks from the base characters. Evaluation of JDasher is continuing.

5. THE LANGUAGE MODEL

When choosing a language model for Dasher we considered two qualities: how well the language model compresses text and the time taken to compute

Figure 13. JDasher–Hiragana character set.

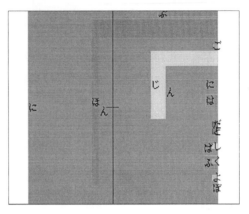

the probability of a character. Dasher must calculate many probabilities each time the screen updates.

The model in Dasher version 1.6 is based on a popular text-compression algorithm called prediction by partial match (PPM; Bell et al., 1990). PPM is a context-based algorithm that uses the preceding characters to predict the next one. The maximum size of the context is the order of the model. We use a variant of the algorithm called PPM5D+ (Teahan, 1995), which is fifth order and can compress most English text to around 2 bpc. There are slightly better algorithms (Gilchrist, 2000), but PPM5D+ is simple and fast.

Given a context, we compute the probabilities of all the symbols in the alphabet. The conditional probabilities determine the intervals in Equation 1.

When using Dasher, it is difficult to select a character that has a very small probability. Therefore, we add a small fixed probability δ to every character and then renormalize the probabilities. In Dasher 1.6, the default value of δ is .002. As the user enters text, the characters can be fed back into the PPM algorithm, hence adjusting the future probabilities in accordance with that user's vocabulary.

6. EMPIRICAL EVALUATION

The following evaluation was carried out on Dasher version 1.3. The results were also presented in Ward et al. (2000).

6.1. Evaluation Approach

An objective of this research was to assess Dasher in a realistic text entry task. Dasher requires uninterrupted visual attention, so we measured the writing speed for dictated text.

Dasher has been used with a variety of position controllers. For the experiment, we originally hoped to use a stylus on a touch screen the size of a handheld computer. However, our original touch screen required a contact force of 2 oz. This is difficult to maintain when using a constant gestural interface such as Dasher, so we used a standard mouse in the experiment described next.

6.2. Pilot Experiment

An initial approach to the evaluation was for the user to transcribe text passages spoken by a speech synthesizer. The speed of speech was dynamically controlled to stay just ahead of the words being entered by the user.

The results of the three-participant pilot experiment were used to estimate an appropriate length for each experimental session, design the session format, and estimate the number of participants and training sessions that would be required to observe significant learning. Participants found the speech synthesizer hard to understand, so we used recorded human speech for the main experiment.

6.3. Method

Participants

The main evaluation experiment involved 10 experimental participants, recruited from the students and staff of the Cavendish Laboratory. All participants had vision corrected to normal and spoke English as their first language. Participants were paid for participating in the experiment. None of the participants had previous experience with Dasher.

Task

The experimental task, as in the pilot experiment, was to enter text dictated from Jane Austen's *Emma* (Project Gutenberg). We selected 18 extracts at random from the total 883 kbytes of text in *Emma* to form the test set. The remaining 866 kbytes of text was used to train the language model.

Austen has a distinctive writing style. This allowed us to test Dasher in a reasonable simulation of typical usage where the language model would have been trained on previous prose written by the same user.

Apparatus

Platform	Pentium II 300 MHz running Linux 2.0.38
Monitor	38-cm LCD display
Input Device	Mouse

The dictated passages were recorded as a series of audio files, with each file containing a short phrase. The text entered by the participant was monitored so that as he or she wrote the penultimate word in each phrase, the next audio file was played automatically. This allowed participants to enter text continuously, with dictation proceeding at a comfortable speed. A simple synchronization algorithm compensated for misspelled words when identifying the end of a dictation phrase. Participants could press a key with their free hand to repeat the last phrase if it was forgotten or misheard. There was no limit to the number of repeats.

Procedure

Participants completed six experimental sessions of approximately 30 min each. Each session consisted of three exercises. Participants first took dictation using Dasher for 5 min. They then took dictation using the keyboard (conventional typing) for 2 min. Finally they took dictation using Dasher for another 5 min. Sessions were generally spaced at daily intervals, with no more than two on any day and no more than 3 days between sessions. Participants were instructed at the start of each exercise to write as fast as possible, and they were told that they could correct simple mistakes within the current word. They were told not to correct mistakes in previous words.

Configuration

Dasher was configured with the following parameters:

Display	600 × 500 pixels, measuring 12 cm × 10 cm
Character set	27 (lowercase and Space)
Minimum threshold for plotting rectangles	4 pixels
δ	.002
Steps parameter, S	Adaptive, initialized to 60

The δ parameter (defined in Section 5) specifies a lower limit to the probability of a character. For each exercise, the Steps parameter S (defined in Section 3.3) was initialized to 60.

Analysis

The control software counted the characters entered in each period of dictation. It also counted word-level errors: a deletion, insertion, or replacement of a word (possibly with a misspelled version of the word)—each counted as an error. The Viterbi algorithm (Bellman, 1957; Viterbi, 1967) was used to determine the minimum number of errors that, when applied to the target text, produced the text entered by the user.

Data on character entry speed and proportion of errors were collected for the 12 periods of Dasher use and for the 6 periods of conventional typing. We compared improvement in performance over the 12 periods of Dasher use to evaluate the effect of practice on Dasher performance. The interpolated conventional typing tests allowed us to estimate what proportion of this practice effect could be attributed to practice with the experimental dictation software, as opposed to practice with the Dasher text entry method.

The keyboard task should be viewed as a control condition for the experimental procedure rather than a serious attempt to measure keyboard speeds under realistic conditions of use.

6.4. Results

Text Entry Speeds

All participants successfully completed the six sessions (Figure 14). The dictation texts were used in the same order for all participants. As a result, variation in vocabulary between the texts resulted in consistently lower speeds for some exercises. For example, Exercise 8 included a particularly uncommon word that caused users to hesitate. As a result, the raw data shown in Figures 14, 17, and 20 include apparent dips in writing speed during those exercises.

To investigate these correlations further, we plot information rate rather than writing speed (Figure 15). The correlations between participants over Exercises 7, 8, and 12 are no longer apparent because uncommon words have a higher information content, and hence information rate is relatively continuous despite discontinuities in writing speed when they are encountered.

Writing Errors

Errors were expressed as the percentage of words written differently from the test text. The rate of keyboard errors in Figure 16 shows that users took a little while to get used to the dictation system (assuming that their typing skill did not improve during the experiment). However, after the third keyboard exercise, the errors in both Dasher and keyboard exercises change very little.

Toward the end of the evaluation the average error rate is 3% when using Dasher and 7% when using the keyboard. On 10 occasions, Dasher users made no errors whereas keyboard users invariably made errors.

Figure 14. Effect of practice on writing speeds for 10 participants. Each exercise was 5 min long.

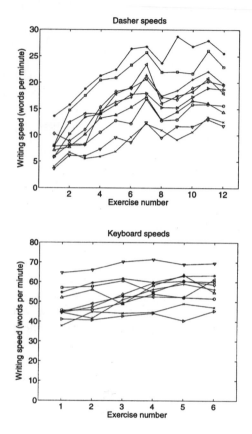

Figure 15. Dasher information rate. Each exercise was 5 min long.

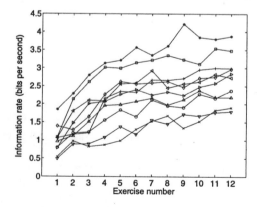

218

Figure 16. Writing errors.

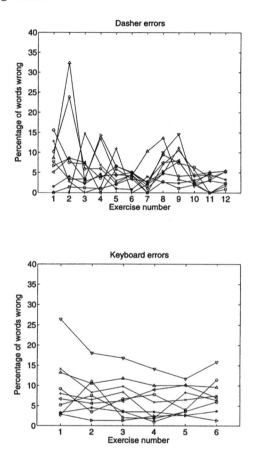

The Power Law of Practice

The power law of practice (Newell & Rosenbloom, 1981) predicts that the writing speed dN/dt should increase with training time t as

$$\log\left[\frac{dN}{dt}\right] = C + D\log(t),$$ (11)

where C and D are constants. From Figure 17 we can see that the data give a reasonably good fit to the power law. The gradient $D = .32 \pm .05$ (related to the learning rate) is similar for all 10 participants, but $C = 3.22 \pm .45$ (the log rate after 1 min) varies more among participants. The users with faster initial speeds were generally the fastest after 1 hr of training. Consequently, we

Figure 17. Logarithmic plot to test the Power Law of Practice. Dasher results are plotted for all 10 participants.

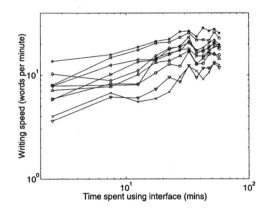

can estimate a users' future typing speed with reasonable accuracy, based on a small amount of use, say, 15 min.

7. INFORMATION RATE OF DASHER

7.1. Current Information Rate of Dasher

Ward has used Dasher for substantially longer than the experimental training period; a total of a few hours' use, mostly during development testing rather than sustained text entry. During a number of trials with the experimental dictation text from Jane Austen's *Emma*, he achieved an average speed of 170 characters per minute (34 wpm).

The experimental texts have an information content of 1.7 bpc with respect to the model. Expert performance of 170 characters per minute therefore corresponds to an information rate of 4.8 bps.

7.2. Potential Information Rate of Dasher

Apart from the language model, two factors might limit the rate at which a user can enter information with Dasher. First, steering Dasher in the required direction involves visual motor control similar to pole-balancing; the timescale of the eye-to-pointer feedback loop imposes a maximum writing rate.

Second, a limit in Dasher might be the time required for the user to search among the presented strings. We performed an experiment to estimate the maximum writing speed of Dasher when this visual search is not required, so only the first factor applies. The required sequence of squares was highlighted in a strongly contrasting color (Figure 18), and the experienced user guided Dasher along this sequence as fast as possible. The required sequence of rectangles and their sizes were identical to those displayed when writing the experimental text from *Emma*. The information content of the required writing path is therefore identical to that of the experimental text.

When operating Dasher as shown in Figure 18, the same user was able to "write" at a rate of 228 characters per minute. This rate corresponds to an information rate of 6.5 bps, which is 1.3 times faster than dictation speed when visual search is required. We conclude that visual search is not a major limit in this implementation.

We now present a quantitative model of the information rate limit imposed by the visual–motor feedback loop. A person is trying to track an object at $y(t)$ with a pointer at $u(t)$. The object runs away from the pointer in accordance with:

$$\frac{d_y}{d_t} = l(y - u), \tag{12}$$

where l is the exponential growth rate of the deviation, which is proportional to the writing rate. We model the person as a tracker with a delay:

$$u(t) = y(t - \tau). \tag{13}$$

Substituting for $u(t)$ in Equation 12,

$$\frac{d_y}{d_t} - ly(t) + ly(t - \tau) = 0. \tag{14}$$

This is a type of delay differential equation with well-known properties (Bellman & Cooke, 1963). Analysis by Laplace transform shows that $y(t)$ is stable if and only if $l\tau < 1$.

The tracking problem is identical to using Dasher, where l is the expansion rate of Dasher, assumed constant. If the visible interval at time $t = 0$ has size 1, then after time t, the visible interval has size

$$p = \exp(-lt). \tag{15}$$

The information content is defined as $-\log_2(p)$, so the information rate of Dasher is

$$\frac{-\log_2(p)}{t} = \frac{l}{\log_e(2)} \text{ bps.} \tag{16}$$

Figure 18. Modified version of Dasher to measure writing speed without visual search.

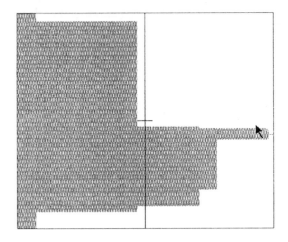

Applying the stability condition $h < 1$, we find an expression for the maximum information rate, M.

$$M = \frac{1}{\tau \log_e(2)} \text{ bps.} \tag{17}$$

If we assume an eye-to-hand reaction time of 175 milliseconds (Rosenbaum, 1991), then the maximum information rate is $M = 8.2$ bps. This theoretical model agrees reasonably well with our measured information rate of 6.5 bps for our most expert user.

In Dasher version 1.0, users could drive the interface arbitrarily fast by holding the mouse near the right-hand vertical. From the analysis of the time-delay model, it would seem appropriate to set an upper limit to the rate of expansion so that the dynamics are always stable. In version 1.6, there is an option to set this maximum information rate.

8. APPLICATION TO MOBILE COMPUTING

We used a variable sized window on a 38-cm LCD touch screen to simulate the use of Dasher on a PDA. This touch screen was not ideal, as it requires an unreasonably large 2 oz of force.

The size of the output display was varied from 50×50 pixels (1.6 cm $\times$ 1.6 cm) to 600×600 pixels (20 cm $\times$ 20 cm). Our screen had 30 pixels per cm, similar to the 27 pixels per cm on the Handspring Visor, a Palm-compatible PDA. For each size, an experienced user performed 10 dictation tasks. We plot the mean writing speed in Figure 19.

Figure 19. Touch screen writing: Effect of screen size.

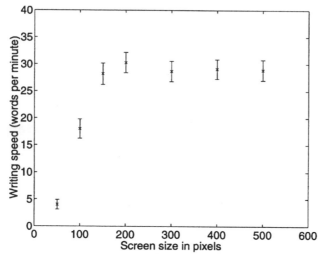

The writing speed decreases rapidly when the screen size is decreased below 150 × 150 pixels. Increasing the screen size beyond 200 × 200 pixels does not appear to be advantageous.

9. DISCUSSION

9.1. Comparison to Other Devices

Where previous researchers have reported improvement of performance over time, we can compare Dasher with other devices.

We have extracted data from a number of earlier published evaluations of text entry interfaces and compare the results to Dasher (Figure 20).

In a relatively short training time, writing performance with Dasher already exceeded that of many methods. If we extrapolate a power law function fit (a straight line on Figure 20), it seems that with further practice, Dasher performance may also equal the fastest method using Qwerty keyboard derivatives.

9.2. Attention Demand

It is clear that Dasher requires sustained visual attention from the user. This requirement is in contrast to conventional keyboards that can be used without visual attention after sufficient training.

Figure 20. Comparison of Dasher to other devices. Hand-printing required the user to print letters on a touch screen. In Qwerty-tapping, a stylus selects characters from an on-screen Qwerty keyboard. ABC-tapping uses a stylus with letters in alphabetical order. Half-Qwerty uses half a Qwerty keyboard, with the other half accessible with a special key. Bellman is a keyboard designed with a probabilistic character layout strategy. TCK2 is a version of the ternary chorded keyboard. OPTI is a soft keyboard with an optimized layout.

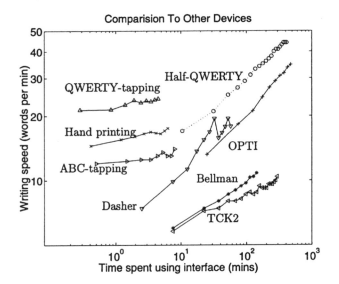

9.3. Potential Writing Speed

The performance of Dasher is determined by two factors. The first is how well the language model can compress the text being entered. Shannon (1948) estimated the entropy of English to be 1 bpc, but PPM typically compresses to 2 bpc. A perfect language model should increase the speed of Dasher by a factor of 2, which would result in a writing speed similar to that of a Qwerty keyboard.

The second factor is the user interface, which can currently convey 4.8 bps to the computer (Section 7.1). Earlier, we cited an upper bound estimate of 8.2 bps for human pointing performance. The following parameters are important:

- Aspect ratio of Dasher and/or the individual rectangles
- Positioning of the characters on the screen
- Pointing device
- Frame-rate of Dasher (averaging 40 frames per second in the experiment)
- Minimum pixel threshold for displaying new rectangles

There is a trade-off between the last two factors; the values will be influenced by the performance of the system.

9.4. Suitability to Mobile Text Entry

The operation of Dasher is performed without a keyboard, using a single continuous stream of motor input. In Section 8 we determined that writing speeds are fairly constant at screen sizes of 150 × 150 and greater. Many PDAs have screens of at least 200 × 200, so we believe that Dasher could be used on a PDA.

Dasher has recently been ported to a PDA. Details on our progress, including a download for Pocket PCs (ARM, MIPS, SH3 processors), can be found at the Dasher Web site http://www.inference.phy.cam.ac.uk.

The combination of stylus and touch screen is well suited to Dasher. The user simply points at the letters that they want to select. This mode of operation should prove to be faster than a mouse, where the user is using an indirect method of pointing.

Dasher can be configured so that the viewpoint moves only while the pen is in contact with the screen. The language model can be trained on an existing dictionary, e-mails, or other documents. As the user writes, new words and phrases can be automatically added to the model.

10. FURTHER WORK

10.1. Eye-Tracking

An eye-tracking device is currently being tested as an alternative input device to Dasher, enabling hands-free text entry. The system could be especially beneficial to users with limited hand movement.

When Dasher is used with an eye-tracker, users simply look at the string they want to write. We hope that this will be easier and more natural to use than eye-tracking interfaces in which the users must consciously direct their gaze toward action targets (e.g., Salcucci & Anderson, 2000).

10.2. Modeling

Since the experimental work, a new language model has been implemented. This model also uses the PPM algorithm, but it is word-based rather than character-based. A dictionary is used to supplement the statistical model. Compression performance on English text is around 10% better than the character model. We observed improvements in typing speed of around 10%.

We are also exploring latent-variable models that should be able to learn new languages more rapidly.

10.3. Japanese

We carried out a small survey of Japanese people to find out the typing speed for Hiragana entered with a Qwerty keyboard. A typical typing rate is about 60 to 120 Hiragana per minute.

We trained a language model on 200k of Japanese text, and we estimated the information content of Hiragana to be 2.5 bits per character. The current rate of entering information with Dasher is about 4.8 bps, so we might expect a writing speed of 2 Hiragana per second or 120 Hiragana per minute.

We hope to perform an experiment to determine the performance of the Hiragana version of Dasher.

11. CONCLUSION

Dasher is a novel text entry interface that exploits the information redundancy of the English language and the human capacity to convey high information rates through fine motor control.

The operation of Dasher is simple and immediately evident to new users. Furthermore, Dasher has a rapid learning rate that is comparable to alternative text entry methods.

We think Dasher shows promise as a keyboard-less text entry system both in its absolute writing speed and ease of use.

NOTES

Acknowledgments. Dasher was conceived by David MacKay and Mike Lewicki and would have not been possible without Ousterhout's Tcl language (Ousterhout, 1994). We thank the experimental participants. Thanks to T. Matsumoto, T. Tokieda, and T. Fujiyoshi for helpful discussions.

Support. This work was supported by the Gatsby Foundation and by the Applied Science Laboratories. Alan Blackwell's research is funded by the Engineering and Physical Sciences Research Council, under EPSRC grant GR/M16924, "New paradigms for visual interaction."

Authors' Present Addresses. David J. Ward, Cavendish Laboratory, Madingley Road, Cambridge, CB3 0HE, United Kingdom. E-mail: djw30@mrao.cam.ac.uk. Alan F. Blackwell, Computer Laboratory, Cambridge, CB3 0FD, United Kingdom. E-mail: Alan.Blackwell@cl.cam.ac.uk. David J.C. MacKay, Cavendish Laboratory, Madingley Road, Cambridge, CB3 0HE, United Kingdom. E-mail: mackay@mrao.cam.ac.uk.

HCI Editorial Record. First manuscript received October 23, 2000. Accepted by Scott Mackenzie. Final manuscript received May 29, 2001. — *Editor*

REFERENCES

Bell, T., Cleary, J., & Witten, I. (1990). *Text compression.* Englewood Cliffs, NJ: Prentice Hall.

Bellman, R. (1957). *Dynamic programming.* Princeton, NJ: Princeton University Press.

Bellman, R., & Cooke, K. L. (1963). *Difference differential equations.* New York: Academic.

Bellman, T., & MacKenzie, I. (1998). A probabilistic character layout strategy for mobile text entry. *Proceedings of Graphics Interface 98.* Toronto: Canadian Information Processing Society.

Borges, J. L. (2000). *The library of Babel.* Boston: David R. Godine.

Gilchrist, G. (2000). *Archive comparison test.* Retrieved from http://compression.ca

Goldberg, D., & Richardson, C. (1993). Touch-typing with a stylus. *Proceedings of the INTERCHI 93 Conference on Human Factors in Computing Systems.* New York: ACM.

Gordon, M., Henry, H., & Massengill, D. (1975). Studies in typewriter keyboard modification: I. Effects of amount of change, finger load, and copy content on accuracy and speed. *Journal of Applied Psychology, 60,* 220–226.

Hunter, M., Zhai, S., & Smith, B. A. (2000). Physics-based graphical keyboard design. *Proceedings of CHI 2000 Conference on Human Factors in Computing Systems.* New York: ACM.

James, C. L., & Reischel, K. M. (2001). Text input for mobile devices: Comparing model prediction to actual performance. *Proceedings of CHI 2001 Conference on Human Factors in Computing Systems.* New York: ACM.

Jecker, D. (1999, November). Benchmark tests: Speech recognition. *PC Magazine.*

Kroemer, K. (1992). Performance on a prototype keyboard with ternary chorded keys. *Applied Ergonomics, 23*(2), 83–90.

MacKenzie, I. (1992a). Fitts' law as a research and design tool in human–computer interaction. *Human–Computer Interaction, 7,* 91–139.

MacKenzie, I. (1992b). Movement time prediction in human–computer interfaces. In R. M. Baecker, W. A. S. Buxton, J. Grudin, & S. Greenberg (Eds.), *Readings in human–computer interaction* (2nd ed., pp. 483–493). Los Altos, CA: Kaufmann.

MacKenzie, I., & Zhang, S. (1997). The immediate usability of Graffiti. *Proceedings of Graphics Interface 97.* Toronto: Canadian Information Processing Society.

MacKenzie, I. S., Zhang, S. X., & Soukoreff, R. W. (1999). Text entry using soft keyboards. *Behaviour & Information Technology, 18,* 235–244.

Matias, E., MacKenzie, I., & Buxton, W. (1993). Half-Qwerty: A one-handed keyboard facilitating skill transfer from Qwerty. *Proceedings of the INTERCHI 93 Conference on Human Factors in Computing Systems.* New York: ACM.

Newell, A., & Rosenbloom, P. (1981). Mechanisms of skill acquisition and the power law of learning. In J. R. Anderson (Ed.), *Cognitive skills and their acquisition* (pp. 1–55). Hillsdale, NJ: Lawrence Erlbaum Associates, Inc.

Norman, D., & Fisher, D. (1982). Why alphabetic keyboards are not easy to use: Keyboard layout doesn't much matter. *Human Factors, 24,* 509–519.

Ousterhout, J. K. (1994). *Tcl and the Tk toolkit.* Reading, MA: Addison-Wesley.

Perlin, K. (1998). Quikwriting: Continuous stylus-based text entry. *Proceedings of the UIST 98 Symposium on User Interface Software and Technology.* New York: ACM.

Rosenbaum, D. A. (1991). *Human motor control.* San Diego, CA: Academic.

Salcucci, D. D., & Anderson, J. R. (2000). Intelligent gaze-added interface. *Proceedings of CHI 2000 Conference on Human Factors in Computer Systems.* New York: ACM.

Sears, A., Revis, D., Swatski, J., Crittenden, R., & Shneiderman, B. (1993). Investigating touchscreen typing: The effect of keyboard size on typing speed. *Behaviour and Information Technology, 12,* 17–22.

Shannon, C. E. (1948). A mathematical theory of communication. *Bell Systems Technical Journal, 27,* 379–423, 623–656.

Silfverberg, M., MacKenzie, I. S., & Korhonen, P. (2000). Predicting text entry speed on mobile phones. *Proceedings of CHI 2000.* New York: ACM.

Teahan, W. (1995). Probability estimation for PPM. *Proceedings NZCSRSC'95.* Retrieved from http://www.cs.waikato.ac.nz/~wjt/papers/NZCSRSC.ps.gz

Venolia, D., & Neiberg, F. (1994). T-Cube: A fast, self-disclosing pen-based alphabet. *Proceedings the CHI 94 Conference on Human Factors in Computer Systems.* New York: ACM.

Viterbi, A. J. (1967). Error bounds for convolutional codes and an asymptotically optimum decoding algorithm. *IEEE Transactions on Information Theory, IT-13.*

Ward, D. J., Blackwell, A. F., & MacKay, D. J. C. (2000). Dasher: A data entry interface using continuous gestures and language models. *Proceedings of UIST 2000.* New York: ACM.

Zhai, S., & Smith, B. (2001). Alphabetically biased virtual keyboards are easier to use—layout does matter. *Proceedings of the CHI 2001 Conference on Human Factors in computer Systems.* New York: ACM.

HUMAN-COMPUTER INTERACTION, 2002, Volume 17, pp. 229–269

Performance Optimization of Virtual Keyboards

Shumin Zhai, Michael Hunter, and Barton A. Smith

IBM Almaden Research Center

ABSTRACT

Text entry has been a bottleneck of nontraditional computing devices. One of the promising methods is the virtual keyboard for touch screens. Correcting previous estimates on virtual keyboard efficiency in the literature, we estimated the potential performance of the existing Qwerty, FITALY, and OPTI designs of virtual keyboards to be in the neighborhood of 28, 36, and 38 words per minute (wpm), respectively. This article presents 2 quantitative design techniques to search for virtual keyboard layouts. The first technique simulated the dynamics of a keyboard with digraph springs between keys, which produced a Hooke keyboard with 41.6 wpm movement efficiency. The second technique used a Metropolis random walk algorithm guided by a "Fitts-digraph energy" objective function that quantifies the movement efficiency of a virtual keyboard. This method produced various Metropolis keyboards with different shapes and

Shumin Zhai is a human–computer interaction researcher with an interest in inventing and analyzing interaction methods and devices based on human performance insights and experimentation; he is a Research Staff Member in the User Sciences and Experience Research Department of the IBM Almaden Research Center. **Michael Hunter** is a graduate student of Computer Science at Brigham Young University; he is interested in designing graphical and haptic user interfaces. **Barton A. Smith** is an experimental scientist with an interest in machines, people, and society; he is manager of the Human Interface Research Group at the IBM Almaden Research Center.

CONTENTS

structures with approximately 42.5 wpm movement efficiency, which was 50% higher than Qwerty and 10% higher than OPTI. With a small reduction (41.16 wpm) of movement efficiency, we introduced 2 more design objectives that produced the ATOMIK layout. One was alphabetical tuning that placed the keys with a tendency from A to Z so a novice user could more easily locate the keys. The other was word connectivity enhancement so the most frequent words were easier to find, remember, and type.

1. INTRODUCTION

Pervasive devices have come to the forefront in computer technology. Small handheld devices such as personal digital assistants (PDAs), pagers, and mobile phones, as well as larger scale devices such as tablet computers and electronic whiteboards, now play an increasingly more central role in human–information interaction. This general trend is rapidly freeing us from the confines of our laptop or desktop computers and leading us to a future of pervasive computing.

Despite this favorable trend in pervasive computing, certain obstacles stand in the way of developing efficient applications on these devices. An obvious problem relates to text input. For example, a recent study by McClard and Somers (2000) clearly demonstrated the value of tablet computers in home environments. However, the lack of efficient text input techniques in these tablet computers made many common applications, such as chat, e-mail, or even entering a URL very difficult.

Text entry is also problematic for PDAs and other handheld devices. Currently, text input on these devices can be achieved through reduced physical keyboards, handwriting recognition, voice recognition, and virtual keyboards, but each has critical usability shortcomings. We briefly describe several representative methods. See MacKenzie and Soukoreff (2002) for a more detailed survey of these methods.

Physical Keyboards. There are two ways to reduce the size of physical keyboards. One is to shrink the size of each key. This is commonly seen in electronic dictionaries. Typing on these keyboards is slow and difficult due to their reduced size. The other method is to use the number pads in telephones, whereby each number corresponds to multiple letters. The ambiguity of multiple possible letters is commonly resolved by the number of consecutive taps, or by lexical models.

Handwriting. Reducing error rate has been the major goal in handwriting recognition. However, the ultimate bottleneck of handwriting input lies in the human handwriting speed limit. It is very difficult to write legibly at a high speed.

Voice Recognition. Speech has been expected to be a compelling alternative to typing. Despite the progress made in speech recognition technology, however, a recent study by Karat, Halverson, Horn, and Karat (1999) showed that the effective speed of text entry by continuous speech recognition was still far lower than that of the keyboard (13.6 vs. 32.5 corrected words per minute [wpm] for transcription and 7.8 vs. 19.0 corrected wpm for composition). Furthermore, the study also revealed many human-factors issues that had not been well understood. For example, many users found it "harder to talk and think than type and think" and considered the keyboard to be more "natural" than speech for text entry.

There have also been other continuous-gesture-based text methods. A recent example is Dasher (Ward, Blackwell, & MacKay, 2000), which uses continuous mouse movement to pass through traces of letters laid out by a predictive language model. Using such a technique is a novel and intriguing experience, but the primary drawback is that the user has to continuously recognize the dynamically arranged letters. The visual recognition task may limit the eventual performance of text entry with such a method.

This article focuses on the design of virtual keyboards. Such a keyboard displays letters and numbers on a touch sensitive screen or surface. To input text, the user presses keys with a finger or stylus. Such a keyboard can be scaled to fit computing devices with varying sizes, particularly small handheld devices. One central issue, however, is the layout of the keys in these keyboards. Due to developers' and users' existing knowledge, the Qwerty layout used in most physical keyboards today has the momentum to become the most likely choice. In fact, some PDA products, such as the Palm™ Pilot, have already used the Qwerty as their virtual keyboard layout.

Unfortunately, the Qwerty layout (see Figure 1) designed by Christopher L. Sholes, Carlos Glidden, and Samuel W. Soule in 1868 is a poor choice for virtual keyboards. This is because the Qwerty keyboard was so arranged that many adjacent letter pairs (digraphs) appear on the opposite sides of the keyboard. The main purpose of this arrangement was to minimize mechanical jamming (Cooper, 1983; Yamada, 1980). Accidentally, this design also facilitates the frequent alternation of the left and right hand, which is a key premise to rapid touch typing with two hands that was discovered many years after the typewriter was invented. Partially because the Qwerty design scores well in alternation frequency, various attempts to replace Qwerty with more efficient layouts, such as the Dvorak simplified keyboard (Dvorak, Merrick, Dealey, & Ford, 1936), have not prevailed. The performance gain with these newer designs (around 15%) has not been substantial enough to justify the cost of retraining the great number of Qwerty users (Cooper, 1983; Norman & Fisher, 1982; Yamada, 1980). However, on a virtual keyboard, the polarizing common digraphs in Qwerty mean that the stylus has to move back and forth more frequently and over greater distances than necessary. The key to a good virtual keyboard is exactly opposite to the idea behind Qwerty. Common digraph letters should be close to each other so the hand does not have to travel much. The movement distance concern also points to another problem of Qwerty as a virtual keyboard layout, it is elongated horizontally, which statistically increases the stylus movement distances. In fact, the human performance effect of relative distances between the letters can be modeled by a simple movement equation—the Fitts' law.

There are additional reasons to thoroughly study virtual keyboard layouts at this point in user interface history. First, it is not too late to form a new layout

Figure 1. The Qwerty layout designed by Sholes, Glidden, and Soule in 1868.

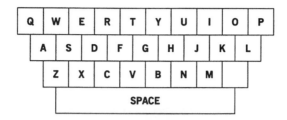

standard for the virtual keyboard, due to the relatively small number of people using virtual keyboards today. Second, the 10-finger touch typing skills on a physical keyboard do not necessarily transfer to on-screen stylus tapping on the same layout (Zhang, 1998). The perceptual, memory, and motor behavior of using a virtual keyboard is sufficiently different from that of a physical keyboard to justify a different design.

2. PERFORMANCE MODELING OF VIRTUAL KEYBOARDS

To minimize finger movement on a virtual keyboard, two factors must be taken into account. One is the transitional frequencies from one letter to another in a given language (digraph statistics), and the other is the relative distances between keys. The goal should be to arrange the letters so that the statistical total travel distance is the shortest when tapping on such a keyboard. This means that the most frequent keys should be located in the center of the keyboard and the frequently connected letters (such as *T* and *H*) should be closer to each other than the less frequently connected letters.

2.1. The Fitts-Digraph Model of Virtual Keyboards

MacKenzie and colleagues (MacKenzie & Zhang, 1999; Soukoreff & MacKenzie, 1995) were the first to use a quantitative approach to model virtual keyboard performance. Their model predicts user performance by summing the Fitts' law movement times (MTs) between all digraphs, weighted by the frequencies of occurrence of the digraphs. The use of Fitts' law made it possible to estimate performance in absolute terms, giving us a comparison to speed we are familiar with, such as 60 wpm for a good touch typist.

According to Fitts' law (Figure 2), the time to move the tapping stylus from one key i to another j for a given distance (D_{ij}) and key size (W_j) is[1]

---

1. i and j here represent any pair of keys from A to Z and the space key.

Figure 2. The average movement time can be predicted by Fitts' law.

$$MT = a + b\, Log_2(D_{ij}/W_j + 1)$$

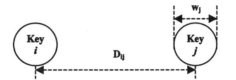

$$MT = a + b \log_2\left(\frac{D_{ij}}{W_j} + 1\right), \tag{1}$$

where *a* and *b* are empirically determined coefficients. To be able to make comparisons to the results in the literature (e.g., MacKenzie & Zhang, 1999), we choose $a = 0$, $b = 1 / 4.9$. In other words, we consider the Fitts' index of performance (IP; Fitts, 1954) to be 4.9 bits per second (bps). We return to the choice of this parameter later.

If the frequency of letter *j* to follow letter *i* (digraph *I–j*) among all digraphs is P_{ij}, then the mean time in seconds for typing a character is:

$$t = \sum_{i=1}^{27}\sum_{j=1}^{27} \frac{P_{ij}}{IP}\left[log_2\left(\frac{D_{ij}}{W_j} + 1\right)\right], \tag{2}$$

Assuming five characters per word (including space key), this equation allows us to calculate tapping speed in wpm (60 / 5 *t*).

Note that a special case has to be made for Equation 2 when $i = j$. This is when the user taps on the same key successively (e.g., *oo* as in *look*). In this case, the second term in Equation 1 is 0 but *a* is set at .127 sec. Previous researchers (MacKenzie & Zhang, 1999; Zhang, 1998) have used both .127 and .135 sec. We chose .127 because it was closer to what we measured (7.8 repeats on the same location). The influence of this number is small, however, due to the low frequency of such cases.

It should be emphasized that Equation 2 only estimates the movement efficiency of tapping on a virtual keyboard. An expert user could possibly achieve this efficiency. A novice or intermediate user has to visually search for the destination key before tapping on it. In that sense, Equation 2 only predicts the potential upper bound of a user's performance (Soukoreff & MacKenzie, 1995). However, the Fitts' law coefficient in the model is based on average human tapping performance. Some users, therefore, could surpass this "upper bound."

2.2. Digraph Frequency

The digraph frequency P_{ij} in Equation 2 is numerically calculated by the ratio between the number of i–j digraphs and the total number of digraphs in an English text corpus. One commonly used digraph table was made by Mayzner and Tresselt, extracted from an English text corpus of 87,296 characters (Mayzner & Tresselt, 1965). There are two shortcomings to this digraph table. First, it is not clear whether their corpus still accurately reflects current English word use, in particular the sort of language used in digital media. Second, their corpus was limited to words with three to seven letters. This restriction eliminated many of the most frequently used words (e.g., *I, in, on, is, at, to, of, it,* and *if*).

We hence constructed two new text corpora and two digraph tables. One was sampled from online news articles from sources such as the *New York Times,* the *LA Times* and the *San Jose Mercury News.* The articles covered a range of topics from technical to social–political. The size of the corpus was 101,468 characters (without counting spaces between the words). The second 1,364,497-character corpus was gathered from logs of six online chat rooms. The names of the arbitrarily selected chat rooms were Teen, Atheism, ChristianDebate, Myecamp, CityoftheGreats, and MaisonlkkoguRPG. The text in the chat room corpus consisted of very informal conversations. Most input strings were less than 80 characters, and many were not complete sentences. They were frequently a one- or two-word answer to a question or comment about a previous statement. Capitalization at the beginning and periods at the end of sentences were frequently missing. We included only those records that appeared to be typed by a person. Computer-generated headers and other text were deleted.

We collected two corpora because of the informal style of English used in applications such as e-mail, instant messaging, and chat, for which a virtual keyboard will likely be used. We were concerned that a keyboard optimized for standard formal English may not be optimal for informal electronic writing. As we see later in this article, however, the difference in language style does not significantly alter the movement efficiency of a virtual keyboard, possibly because the phonology of the English language, which determines the digraph distribution, does not change significantly with the formality of the language.

The set of characters tabulated is also different from the digraph table of Mayzner and Tresselt (1965). Their table was concerned only with alphabetical characters. In fact, designers had to reconstruct the Mayzner and Tresselt table with an added space key by inference (e.g., Soukoreff & MacKenzie, 1995). We collected and tabulated 128 × 128 symbols, including many nonalphabetic characters. Because of space limitations, we could provide only the core portions of the digraph tables in the Appendix. They included the space, Roman alphabet, and nine most frequent punctuation mark characters in each of the corpora.[2] In tabulating these data, heading text generated by the IRC program was first deleted, then uppercase Roman alphabet characters were converted to lowercase before the digraphs were counted. Figures 3 and 4 illustrate the most frequent symbols in the two corpora.

To be able to compare with data reported in the literature, we continued to use the Mayzner and Tresselt table in evaluating existing keyboard designs. We used all three tables in designing the Metropolis keyboards presented later in this article.

2.3. Existing Layouts and Their Movement Efficiency Estimation

To put in context our proposed virtual keyboard design (presented in a later section), this section details various existing layouts of virtual keyboards and applies the Fitts-digraph model (Equation 2) to estimate the movement efficiency of these layouts. This is necessary for two reasons. First, other than informal arguments and promotional data, some of the existing layouts have never been scientifically evaluated. Second, previously published estimates of Qwerty and OPTI keyboards (MacKenzie & Zhang, 1999; Zhang, 1998) in the literature need to be corrected.

Qwerty

Applying Equation 2 to evaluate the movement efficiency of the Qwerty keyboard is straightforward, except with respect to the treatment of the Space key. The Space key in the Qwerty layout has a much greater length than the rest of the keys. The Fitts' law distance between a character key and the Space key varies depending on what point of the Space key is tapped. MacKenzie and Zhang (MacKenzie & Zhang, 1999; Zhang, 1998) used a "suboptimal" model to handle the Space key, which divided the Space key into multiple segments; each was equal in length to a regular character key. For each character–space–character "trigraph," they chose the segment of the Space key that yielded the shortest total distance of character–space–character path. Then they calculated the two Fitts' law times of character–space and of space–character according to the distances along that path. Finally, they summed the Fitts' tapping times of all of character–space–character trigraphs, weighted by the probability of each trigraph occurrence. Using this approach, they estimated the Qwerty movement efficiency of 43.2 wpm.

---

2. We will provide the complete tables to researchers upon request.

Figure 3. Letter frequency in the chat room corpus.

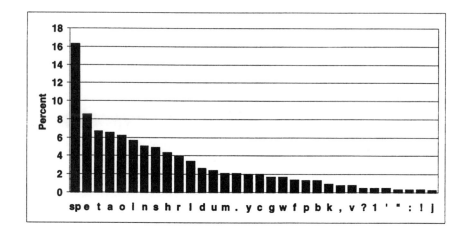

Figure 4. Letter frequency in the news corpus.

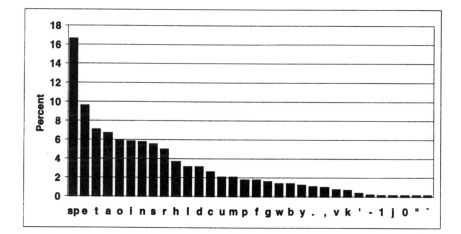

Following the same methodology, we could not replicate their result. On close examination, we found a subtle error in MacKenzie and Zhang's calculation of the probability of each character–space–character trigraph. Taking the combination of i–space–j as an example, they incorrectly used $P_{i\text{-}space} \times P_{space\text{-}j}$ to calculate the probability of the path $P_{i\text{-}space\text{-}j}$. Note that $P_{space\text{-}j}$ is the probability (or frequency) of the transition at any given tapping to be from the space key to the J key. It is not the *conditional* probability of $P_{space\text{-}j/space}$ $(P_{space\text{-}j} / P_{space})$

needed to calculate the probability of two serial events. The correct calculation should be:

$$P_{i\text{-}space\text{-}j} = P_{i\text{-}space} \times P_{space\text{-}j/space} = P_{i\text{-}space} \times P_{space\text{-}j} / P_{space} \qquad (3)$$

where $P_{i\text{-}space}$ is the probability of i-space digraph at any given tapping, $P_{space\text{-}j}$ is the probability of space–j at any given tapping, $P_{space\text{-}j/space}$ is the conditional probability of space–j, given the last tapped key is space, and P_{space} is the probability (frequency) of the Space key.

Using this corrected $P_{space\text{-}j/space}$ calculation but following the same "suboptimal" methodology as in (Zhang, 1998), we found the movement efficiency of Qwerty layout to be 30.5 wpm.

To ensure the correctness of our estimation, we also applied two much simpler methods, one conservative and one liberal. Both involved only digraphs and avoided character–space–character trigraphs. By the conservative method, character-to-space distance was always measured to the center of the Space key. Obviously the result of this should be lower than the estimation of suboptimal model. Indeed, the speed calculated by this method was 27.6 wpm. By the liberal method, the distance between the Space key and any character key was always measured along the shortest (vertical) line from the character to the Space key. Due to the "free warping" effect—the stylus goes into the Space key from one point and comes out from another point of the Space key without taking any time—this should produce a higher estimate than the suboptimal model. Indeed, we found the tapping speed to be 31.77 wpm with this method.

In conclusion, the movement efficiency of a Qwerty keyboard is about 30 wpm, assuming the user always taps on the portion of the Space key that minimizes the character–space–character total path. If the user does not plan one key ahead, the movement efficiency would be about 28 wpm.

Square Alphabetic

We have pointed out that Qwerty is worse than a random design as a virtual keyboard due to the polarization of common digraphs and its elongated shape. As a comparison to Qwerty, we evaluated a design that sequentially lays out the alphabetical letters in a 5 × 6 (column × row) grid (Figure 5). In fact, this was one of the layouts suggested by Lewis, Kennedy, and LaLomia (1999). Although this "design" does not consider digraph frequency distribution at all, its speed is 33.45 wpm, still faster than Qwerty. Note that such a result is based on the conservative assumption that the user also taps the middle of the spacebar at the bottom of the keyboard. Because the Space key is the most frequent key used, an obvious improvement is a 6 × 5 (column × row) design, as in Figure 6.

Figure 5. A square alphabetic layout.

A	B	C	D	E
F	G	H	I	J
K	L	M	N	O
P	Q	R	S	T
U	V	W	X	Y
Z	SPACE			

Figure 6. An improved alphabetic layout.

A	B	C	D	E	F
G	H	I	J	K	L
M	N	O	P	Q	R
S	T	U	V	W	X
Y	Z	SPACE			

MacKenzie–Zhang OPTI

MacKenzie and Zhang (MacKenzie & Zhang, 1999; Zhang, 1998) designed a new, optimized layout, dubbed OPTI (Figure 7). They first placed the 10 most frequent letters in the center of the keyboard. Then, assigning the 10 most frequent digraphs to the top 10 keys, they placed the remaining letters. The placement was all done by trial and error. They later made a further improved 5 × 6 layout, shown in Figure 8. For convenience, we call the 5 × 6 layout OPTI II in this article.

There are four space keys in both OPTI keyboards, evenly distributed in the layout. The user is free to choose any one of them. The optimal choice depends on both the preceding and following key to the Space key. For example, for the sequence of *M*–space–*V* (Figure 8), the upper right Space key is the best choice. However, the upper right space key is not the optimal choice if the tapping sequence is *M*–space–*Y*. In practice, the use of the optimal Space key ranged from 38% to 47%, depending on the user's experience (MacKenzie & Zhang, 1999).

Assuming optimal choice of Space keys, MacKenzie and Zhang (MacKenzie & Zhang, 1999; Zhang, 1998) predicted 58.2 wpm movement efficiency of the OPTI keyboard and 59.4 wpm on the OPTI II layout. These were surpris-

Figure 7. MacKenzie's and Zhang's OPTI layout.

Q	F	U	M	C	K	Z
		O	T	H		
B	S	R	E	A	W	X
		I	N	D		
J	P	V	G	L	Y	

Figure 8. The improved OPTI layout in a 5 × 6 layout (OPTI II).

Q	K	C	G	V	J
	S	I	N	D	
W	T	H	E	A	M
	U	O	R	L	
Z	B	F	Y	P	X

ingly high performance scores that we could not replicate. As in their Qwerty estimation, MacKenzie and Zhang (MacKenzie & Zhang, 1999; Zhang, 1998) used the character–space–character trigraph approach to handle the multiple Space keys. They made the same probability miscalculation of the trigraphs on the OPTIs as they did on the Qwerty.

We recalculated the movement efficiency of the OPTI II keyboard. The first approach was the same as that of MacKenzie and Zhang, except we used the corrected conditional probability in calculating trigraph probabilities. Our result for the OPTI II is 40.3 wpm. This result was based on the assumption that the user *always* used the optimal Space key.

Our second approach took the nonoptimal Space keys into account. We assumed that the user would make use of the optimal Space key 50% of time, which was still higher than the highest actual rate measured (MacKenzie & Zhang, 1999). For the rest of the time, the average distance from the character key to the three nonoptimal Space keys was used. Using this approach, we found the OPTI II movement efficiency to be 36 wpm.

In conclusion, the OPTI II movement efficiency should be between 36 and 40.3 wpm, depending on the optimality of the Space key choice. If we take 38 wpm as a fair (but optimistic) estimate, this is a 35% improvement over Qwerty (28 wpm).

MacKenzie and Zhang conducted an experiment to investigate how quickly users could reach the predicted movement efficiency. In their test, participants reached 44.3 wpm after 20 sessions of text entry, each for 45 min, on the OPTI design (MacKenzie & Zhang, 1999). This is higher than our predicted performance of 38 wpm. We think this disparity is due to the low value of Fitts' law IP used in Equation 2. Originated in MacKenzie, Sellen, and Buxton (1991), 4.9 bps might be an overly conservative estimate of human tapping performance.[3] For two adjacent keys (1 bit), 4.9 bps means 204 ms per tap. This is much longer than what we measured (average 160 ms). The rate of IP reported in the literature for tapping varied widely. For example, Fitts' original report was 10.6 bps (Fitts, 1954). This rate dropped to 8.2 bps after adjusting for the effective width and for the Shannon–MacKenzie formulation (MacKenzie, 1992), but it is still much higher than 4.9 bps. In this article we continue to use 4.9 bps as the default value of IP for two reasons. First, it is on the conservative side. Second, results based on this assumption can be compared with data from previous studies. One should be aware, however, that all predicted performance scores in this article can be proportionally scaled according to the IP rate. For example, if we use 6 bps instead of 4.9 bps, the OPTI II movement efficiency would be $38 \times 6/4.9 = 46.5$ wpm. Figure 9 lists our movement efficiency estimates of various layouts when IP is 4.9, 6, and 8 bps.

FITALY

The FITALY keyboard (Figure 10) is a commercial product by Textware™ Solutions. The design rationale behind this layout included center placement of more frequent keys, dual double-sized Space keys, and the consideration of digraph frequencies (Textware Solutions, 1998).

In a loosely controlled contest (self-reporting with a witness, best performer rewarded with a prize), Textware Solutions collected 34 entries of text entry speed on a Palm Pilot PDA, with 19 contestants using FITALY, 9 using Graffiti, and 6 using Qwerty. FITALY received the highest average score (44.4 wpm), followed by Qwerty keyboard and Graffiti handwriting (both 28.2 wpm; Textware Solutions, 1998). Note that these scores were collected from motivated contestants.

Applying the Fitts-digraph movement efficiency model (Equation 2), we did a formal estimation of the FITALY layout. In our calculation, the two double-sized Space keys were treated differently from other regular keys. First, the

---

3. For this reason, Soukoreff and MacKenzie (1995) also duplicated their estimation with 14 bps.

Figure 9. Summary of movement efficiency (in words per minute [wpm]) of virtual keyboards at different Fitts' law Index of Performance (IP) levels.

Layouts	IP = 4.9 bits/sec	IP = 6.0 bits/sec	IP = 8.0 bits/sec	Estimation Bias
Qwerty	28	34.3	45.7	Slightly conservative[a]
5 × 6 Alphabetic	33.5	41.02	54.7	Slightly conservative[a]
OPTI II (MacKenzie–Zhang)	38	46.5	62.0	Liberal[b]
Fitaly (Textware Solutions)	36	44.1	58.8	Liberal[c]
Lewis–Kennedy–LaLomia	37.1	45.4	60.6	Slightly conservative[a]
Hooke (Zhai–Hunter–Smith)	41.6	50.9	67.9	Slightly conservative[d]
Metropolis (Zhai–Hunter–Smith)	43.1	52.1	69.4	Slightly conservative[d]
ATOMIK (Zhai–Hunter–Smith)	41.2	50.4	67.2	Slightly conservative[d]

[a]Assuming the user always taps the center of the long Space bar. All calculations in this table are based on Mayzner and Tresselt digraph statistics. [b]The performance estimation of the OPTI design changes significantly with the percentage of optimal choice of the Space keys, which requires looking ahead of the current character being tapped. If the user makes optimal choice of the four Space keys less than 50% of the time, OPTI performance drops to 36 wpm. [c]If the user makes optimal choice of the two Space keys less than 75% of the time, Fitaly performance drops to 35.2 wpm. Further, the Space bar is calculated twice as wide as other keys. This is only true to lateral movement. [d]The diameter of the inscribed circle of the hexagon keys was used as the target size. Efficiency drops slightly if square keys are used. For example, at 4.9 bits/sec, the ATOMIK layout (Figure 23) changes from 41.2 to 39.9 wpm in case of square keys.

Figure 10. The FITALY keyboard.

Z	V	C	H	W	K
F	I	T	A	L	Y
		N	E		
G	D	O	R	S	B
Q	J	U	M	P	X

width of the Space keys was considered twice the size of a regular key in the Fitts' law calculation. This clearly was an overestimate when the movement was primarily vertical. Second, there was again the issue of which Space key to use in calculating distances. Two methods were used to deal with this issue. The first always used the closest Space key to each character, with "free warping" between the two Space keys. By this method, the FITALY keyboard performance was estimated as 37.07 wpm. The second method used the shortest character–space distance 75% of the time. The rest of the time the farther Space key was used in calculating distance. By this method, performance of 35.2 wpm was found.

In summary, the movement efficiency of the FITALY keyboard is about 36 wpm, far more efficient than Qwerty, as the company advertised, but less efficient than OPTI II.

Chubon

Figure 11 shows the Chubon keyboard layout. Using the same approach as in the case of the Qwerty layout, we estimated its movement efficiency to be 33.3 wpm, assuming free warping. If we assume the user always taps at the center of the Space key, the movement efficiency will be 32 wpm. Both estimates are slower than OPTI and FITALY but still faster than Qwerty.

Lewis–Kennedy–LaLomia

Instead of using various heuristics to generate a layout, Lewis, Kennedy, and LaLomia (1999) used a more systematic method in their design process. They first created a symmetrical matrix of the relative frequency of unordered English-language digraphs and then analyzed this matrix with a "Pathfinder network-definition program" to create a minimally connected network, which formed the basis for their design. Because the method does not consider all digraph connections, one cannot expect a truly optimized layout from such an approach.

The same issue of location and size with regard to the spacebar exists in evaluating Lewis et al's design (Figure 12). We use the midpoint of the spacebar, a conservative approach, in our evaluation. The result is that the Lewis–Kennedy–LaLomia layout speed is 37.14 wpm.

Lewis, LaLomia, and Kennedy (1999) also referenced an earlier design—the modified Getschow, Rosen, and Goodenough-Trepagnier (1986) layout (Figure 13). Our analysis shows that such a design has a speed of 37.8 wpm.

3. DEVELOPING QUANTITATIVE DESIGN METHODS

We have reviewed a few virtual keyboard alternatives to the Qwerty layout. The layout that produces the highest movement efficiency is MacKenzie and Zhang's OPTI II keyboard. Is the OPTI II keyboard the optimal virtual keyboard design? Can we design a virtual keyboard that facilitates higher movement efficiency? Most of the existing designs, at least in part, are based on

Figure 11. The Chubon keyboard.

		V	U	P			
	Q	M	I	T	S	C	K	Z
J	G	N	R	E	H	B	Y	X
		F	O	A	D	L	W	
				SPACE			

Figure 12. The Lewis–Kennedy–LaLomia layout.

Q	R	W	X	Y	
L	U	A	O	F	
T	H	E	N	G	
V	D	I	S	P	
B	C	M	J	K	Z

Figure 13. The revised Getschow–Rosen–Goodenough–Trepagnier Layout by Lewis, LaLomia, and Kennedy (1999).

F	Q	U	S	P	
C	O	T	H	M	
G	I	E	W	X	
K	N	A	R	B	
J	D	L	Y	V	Z

manual trial-and-error approaches, with the help of letter and digraph frequency tables or a minimally connected digraph network. Given the great number of possibilities, human manual exploration can only try out a small fraction of arrangements.

Getschow et al. (1986) introduced one algorithmic approach to virtual keyboard design. They used a "simple assignment procedure called greedy algorithm" that placed alphabetical letters in the most easily accessible positions according to the letter's frequency rank order (pp. 396–398). As the authors stated, the greedy algorithm ignores many arrangements that could be substantially better because it does not consider the letter placement with respect to each other.

The opposite approach to the simple greedy algorithm is exhaustive algorithmic searching that calculates the efficiency of each and every combination of letter arrangement. However, the complexity of that search—$O(n!)$—is approximately 10^{28}, a number too large even for modern computing.

Between the two extremes, we designed and implemented two systematic, physics-based techniques to search for the optimal virtual keyboard. In this section we present the methodologies and results of these two techniques.

3.1. Dynamic Simulation Method

As shown earlier, the goal of good virtual keyboard design is to minimize the statistical travel distance between characters. The more frequent digraphs should be closer together than less frequent digraphs. To achieve this goal, we first designed a dynamic system technique. Imagine a spring connecting every pair of the 27 keys whose initial positions were randomly placed with spaces between the keys. The elasticity of each spring, when turned on, was proportional to the transitional probability between the two keys so that keys with higher transitional probability would be pulled together with greater force. In addition, there is viscous friction between the circle-shaped keys and between the keys and their environment. The steady state when all keys are pulled together forms a candidate virtual keyboard design. Figure 14 illustrates one part of this dynamic system model.

Fortunately, we did not need to build physical models to create the spring–viscosity–mass dynamic systems. Instead we used a mechanical simulation package (Working Model) to simulate it. In the simulation, the springs were "virtual." They did not stop other objects passing through them, hence preventing the springs from being tangled.

The final positions of the keys might still not be at the minimum tension state because some keys could block others from entering a lower energy state. Two methods were used to reduce the deadlock or local minimum states. First, we experimented with different initial states, which had a very significant impact to the end result. Second, each spring had an extended segment (a strut) that held the keys apart so other keys could be pulled through these gaps to reach a lower level of tension. The length of this segment was manually adjusted in the dynamic simulation process. At the end of each simulation cycle, we reduced the length of the adjustable struts to zero so all the keys were pulled against each other, forming a layout of a virtual keyboard. The movement efficiency of the design was then calculated according to Equation 2 and compared with known results. When unsatisfactory, the layout could be "stretched" out to serve as another initial state for the next iteration of the same process. The iteration was repeated until a satisfactory layout was formed. Figure 15 shows the most efficient layout we achieved with this approach. To capture the gist of the spring simula-

Figure 14. (Part of) the dynamic simulation model: Frequent digraphs are connected with stronger springs.

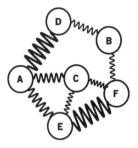

Figure 15. Hooke's keyboard.

tion technique, we call it Hooke's keyboard (after Hooke's Law). The movement efficiency of the Hooke's keyboard shown in Figure 15 is 41.6 wpm, higher than the most efficient previous design (OPTI II, 38 wpm).

3.2. Fitts-Digraph Energy and the Metropolis Method

The idea of minimizing energy, or tension, in the keyboard layout brought us to explore a better known optimization method—the Metropolis algorithm. The Metropolis algorithm is a Monte Carlo method widely used in searching for the minimum energy state in statistical physics (Binder & Heermann, 1988; Metropolis, Rosenbluth, Rosenbluth, Teller, & Teller, 1953; see Beichl & Sullivan, 2000, for a recent review of the Metropolis algorithm). If we define Equation 2 as "Fitts-Digraph energy," the problem of designing a high-performance keyboard is equivalent to searching for the structure of a molecule (the keyboard) at a stable low energy state determined by the interactions among all the atoms (keys). Applying this approach, we designed and implemented a software system that did a "random walk" in the virtual keyboard design space. In each step of the walk, the algorithm picked a key and moved it in a random direction by a random amount to reach a new configuration. The level of Fitts' energy in the new configuration, based on Equation 2, was then evaluated. Whether the new configuration was kept as the starting position for the next iteration depended on the following Metropolis function:

$$W(O - N) = e^{-\Delta E/kT} \quad if \ \Delta E > 0$$
$$= 1 \qquad\quad if \ \Delta E \leq 0$$

(4)

In Equation 4, $W(O - N)$ was the probability of changing from configuration O (old) to configuration N (new), ΔE was the energy change, k was a coefficient, and T was "temperature," which could be interactively adjusted. The use of Equation 4 makes the Metropolis method superior to our previous spring model because the search does not always move toward a local minimum. It occasionally allowed moves with positive energy change to be able to climb out of a local energy minimum.

Again, the initial state where the random walk starts from had a significant impact on the search process. An existing good layout stretched over a larger space was used as an initial state.

In addition to the automatic random walk process itself, we also applied interactive "annealing," as commonly used in the Metropolis searching process. The annealing process involved bringing temperature (T in Equation 4) through several up and down cycles. When temperature was brought up, the system had a higher probability of moving upward in energy and jumping out of local minima. When temperature was brought down, the system descended down to a lower energy level. This annealing process was repeated until no further improvement was seen. Figures 16, 17, and 18 are snapshots from the Metropolis random walk process in one annealing cycle.

We call layouts produced by this process *Metropolis keyboards*. Various layouts with similar movement efficiency were produced. One of them is shown in Figure 19, in which we have replaced the circle shapes used in the design process with hexagons. Each hexagon encapsulated the circle it replaced and filled the gaps between the circles, making more efficient use of the total space. Applying Equation 2, we calculated the movement efficiency of this layout at 42.1 wpm,[4] using the Mayzner and Tresselt table. This is a 50% improvement over Qwerty and more than 10% improvement over OPTI II.

---

4. The diameter of the inscribed circle of a hexagon was used in our Fitts' law calculation. Hence, this is a slightly conservative estimate. If we use the average of diameters of the inscribed and circumscribed circles of a hexagon, the performance here will be 43.7 wpm.

Figure 16. Early stage of Metropolis random walk. The system is at a high energy state, moving toward a lower energy state.

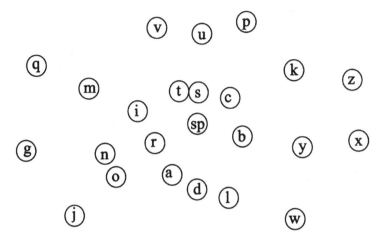

Figure 17. Middle stage of Metropolis random walk. Keys have been descended to a lower energy state. They are getting packed.

Figure 18. Later stage of Metropolis random walk. Keys moved to a lower still energy state.

When we recalculated the speed of the layout in Figure 19 by our chat and news digraph table, we found 41.63 wpm performance by the chat digraph table, and 41.37 wpm by the news table, respectively. The similarity of movement efficiency measured by different corpora is somewhat surprising, given the very different language styles the corpora represent. We initially thought we might have to design virtual keyboards specifically for each type of application, but this is clearly not necessary given the small performance differences. One explanation for the insensitivity of the layout to the corpus is that the language style (formal vs. informal) does not significantly affect the phonology of a corpus, which dictates the digraph distributions. In other words, as long as the language sounds English, the digraph distributions should remain similar. A different language with distinct phonology may indeed require a different layout, although the methods presented here still apply.

The layout shown in Figure 19 does not form a rectangle shape. To make it a rectangle, keys on the outside can be rearranged without great impact on the movement efficiency due to their low frequency. For example, if we move the *J* key to the left of *F* key and *B* key to the left of *D* key, the total movement efficiency of the Metropolis keyboard in Figure 19 would only decrease to 41.6 wpm (using the Mayzner and Tresselt table).

Alternatively, instead of performing the random walk algorithm on an open space where each walk step was taken by moving a randomly selected key to a random direction by a random distance, we could and indeed have developed a program that uses the Metropolis method in a confined array of hexagons. Each walk step was taken by swapping a random pair of keys. Whether such a step took hold depended on the Metropolis function (Equation 4). The rest of the random walk process was the same as the previous approach. Our experience shows that this is an equally effective and more efficient approach, which was used to produce the layouts presented in the rest of the article.

3.3. A Variety of Layouts

By means of the Metropolis algorithm, we designed a variety of layouts with different characteristics, all with similar movement efficiency. This means that it is possible to accommodate design considerations other than the Fitts-digraph energy function. Figure 20 is a layout we accidentally discovered. A particularly interesting characteristic of this layout is that the vowels are connected symmetrically, equally dividing the keyboard into three regions and making the layout more structured. The movement efficiency of this layout is 43.1 wpm, based on the Mayzner and Tresselt digraph statistics. When we applied our digraph tables constructed from current news and chat room text corpora, the movement efficiency was 42.2 wpm and 42.3 wpm, re-

Figure 19. Metropolis Layout 1, with movement efficiency of 42.1, 41.63, and 41.37 wpm based on the Mayzner and Tresselt, chat, and news corpora, respectively.

spectively. Whether this more structured layout is beneficial requires future research.

The Metropolis method can also be used to design a virtual keyboard with any shape desired. All we need to do is lay out the keys in that shape and then let the random walk (swapping) process take over the optimization. Figure 21 gives a triangle example, with 42.45 wpm movement efficiency.

4. ALPHABETICAL TUNING AND WORD CONNECTIVITY: THE ATOMIK LAYOUT

4.1. Alphabetical Tuning

Given the flexibility of producing a variety of layouts with the Metropolis method, we decided to introduce additional characteristics to a layout that may benefit users' learning experience. For novice users of virtual keyboards, speed is determined mostly by the needs to search and find target keys rather than by the amount of motor movement. A keyboard optimized by movement efficiency only may look rather arbitrary to a novice user and hence be difficult to search. We explored the possibility of easing the novice user's search process by introducing alphabetical ordering to a virtual keyboard layout. Alphabetical layout is not a new idea. For example, Norman and Fisher studied a strictly alphabetical layout of the physical keyboard of a typewriter (Norman & Fisher, 1982). They expected, but did not find, that novice users typed faster on such a keyboard than on a standard Qwerty keyboard. The main problem with an alphabetical keyboard, they concluded, was that the keys were laid out sequentially in multiple rows. The location of a key depended on the length of

Figure 20. Metropolis Layout 2, with aligned vowels that divide the keyboard to three regions. Its movement efficiency is 43.1, 42.2, and 42.3 wpm based on the Mayzner and Tresselt, chat, and news corpora, respectively.

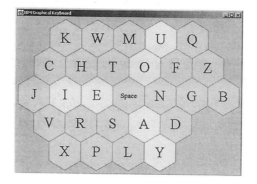

Figure 21. A triangle-shaped layout with 42.45 wpm movement efficiency.

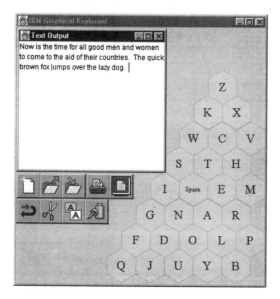

each row—the break point from which the next letter had to start at the left end of the keyboard again. MacKenzie, Zhang, and Soukoreff (1999) studied a virtual keyboard where the letters were laid alphabetically in two columns. Again, they did not find a performance advantage with the alphabetical layout. A confounding factor that might have diminished typing speed in this study was the elongated two-column shape that required an increased average MT. Lewis, LaLomia, et al. (1999) proposed a 5 × 6 virtual keyboard layout with a strictly alphabetical sequence (see Figures 5 and 6). Such a design should suffer from the same problem as discovered by Norman and Fisher—the alphabetical discontinuity caused by row breaks.

Instead of strictly laying out the keys in an alphabetical sequence, we introduce *alphabetical tuning* in the optimization process. To produce such a keyboard, an additional term was added to the "energy" function, which, for each key, depended on the place in the alphabet for the character and on its position on the keyboard:

$$(5)$$

$$e = t + \lambda \sum_{i=a}^{z} \eta(i)(y_i - x_i),$$

where t was the previous energy term defined by Equation 2. λ was an empirically adjusted weighting coefficient, depending on how much alphabetical order was brought to consideration at the cost of the average MT. $\eta(i)$ was an integer number representing the place of the letter in the alphabet, with $\eta(a) = -12$, $\eta(b) = -11$, ... $\eta(m) = 0$, $\eta(n) = 1$, ... and $\eta(z) = 13$. x_i and y_i were the coordinates of letter with origin $(0, 0)$ at the center of the keyboard. The term $\eta(i)(y_i - x_i)$ can be viewed as two forces. $\eta(i)y_i$ produced a force pushing the first half of the letters (a to m) upward and the second half (n to z) downward, with a resulting energy proportional to letter positions. For example, for letter a, $\eta(a) = -12$. The lowest energy state for it is the uppermost position and the highest energy state lies in the lowermost position (negative). The opposite is true for letter z, $\eta(z) = 13$. Similarly, the other force, $\eta(i)(-x_i)$ pushes the first half the letters (a to m) leftward and the latter half rightward. For the Space key, a special case in this treatment, the alphabetic bias term was zero at the center of the keyboard and increases exponentially with distance from the center.

The result of Equation 5 as an objective function was the general trend of letters starting out from the upper left corner moving toward the lower right corner. Figure 22 shows one example of such a design. Clearly, the general tendency of alphabetical order was preserved by this approach, without a significant sacrifice of movement efficiency (41.8 wpm for chat and 41.7 for news).

We conducted a study to test if alphabetical tuning indeed helps novice users. Here we only report a brief summary of the novice user study and refer the reader to Smith and Zhai (2001) for more details. In the study, 12 users with no prior experience with virtual keyboards participated in an order-balanced within-subject experiment with the two layouts, one with alphabetical tuning (shown in Figure 22) and one without alphabetical tuning. The two layouts shared exactly the same geometry (shape and size). With each layout, participants first tapped from the a to the z key as a brief warmup. For the next 15

Figure 22. An alphabetically tuned layout. The first letters tend to appear in the upper left corner, and the last letters tend to appear in the lower right corner.

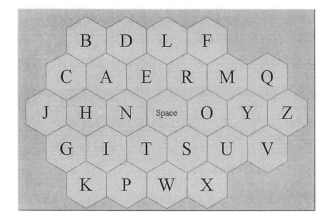

min, they entered memorable English sentences, such as "the quick brown fox jumps over the lazy dog," "we hold these truths to be self-evident," and "all men are created equal." Results show that participants' average speed was 9.7 wpm on the keyboard with alphabetical ordering and 8.9 wpm on the keyboard without alphabetical ordering. The speed difference (9%) between the two conditions was statistically significant, $F(1, 11) = 6.74$, $p < .05$. The error rates were 2% with the alphabetical order and 2.2% without the alphabetical order, $F(1, 11) = .55$, $p = .47$, ns.

To explain the empirical findings of alphabetical tuning, Smith and Zhai (2001) also conducted a theoretical analysis of the uncertainty in visual search, in the framework of the Hick–Hyman law (see Keele, 1986, for a review). With the alphabetical tuning, novices may have a stronger expectation of the area that a letter is likely to appear (lower entropy), hence reduce their search time. This is particularly true for the first ($a, b, c, d, e …$) and last letters (u, v, w, x, y, z).

We call layouts generated with this approach *alphabetically tuned and optimized mobile interface keyboard* (ATOMIK) layouts to reflect both the efficiency and the alphabetical tendency characteristics, as well as the method by which they were produced—atomic interactions between the keys.

4.2. Connectivity of Frequent Words

In using the various layouts produced thus far, we found that the connectivity of a word—the degree to which consecutive letters in the word are adjacent—is very important to movement efficiency, visual search, and memory of the pattern of the word. These motor, visual, and cognitive benefits were not fully characterized by the Fitts-digraph energy. This is particular important to the most common words such as *the*. If a user frequently types a word that is tightly connected, it may ease the user's effort, both real and perceived, hence enhancing the usability and the initial subjective acceptability of virtual keyboards.

We therefore introduced the third criterion in our design—Connectivity Index (CI). CI was defined as

$$CI = \sum_{i=1}^{N} F(i)c(i), \tag{6}$$

where $f(i)$ is the percentage frequency of the ith most frequent word and $c(i)$ is the connectivity score of that word. For example, for the word *the*, if $t-h$ and $h-e$ are connected (adjacent), the word *the* gets a score of 1: $c(i) = 1$. It is multiplied by $f(i) = 3.38\%$, its frequency, before being added to CI. If only $t-h$ or only $h-e$ are connected, the word *the* gets a score of $c(i) = .5$.

In consideration of the Zipf's law[5] effect, we only used the most frequent words to compute CI. These top-ranked words cover a disproportional amount of usage (Figure 23). Excluding single letter words *I* and *a*, we chose the top 17 most frequent words in the chat corpus to compute CI. They were *the, to, you, and, of, is, that, in, it, no, me, are, with, have, was, for,* and *what.*

Figure 24 shows an ATOMIK layout that well satisfies all three criteria—movement efficiency, alphabetical tuning, and word connectivity. The movement efficiency is 41.67 wpm, based on the Mayzner and Tresselt digraph statistics (Mayzner & Tresselt, 1965); 41.16 wpm, based on the chat digraph statistics (Appendix Figure A–1); and 40.78 wpm, based on the news digraph statistics (Appendix Figure A–2).

Because many of the common words are totally connected (e.g., *the, to, and, is,* and *in*), experienced users might be able to stroke through these letters instead of tapping on each of them. Such a strategy may not only save time but also enrich the set of input gestures.

5. ADDITONAL OPTIMIZATION ISSUES

5.1. Auxiliary Keys

Auxiliary keys include all nonalphabetic keys, such as the punctuation keys. How should these auxiliary keys be arranged? Should we optimize them

---

5. Zipf's law models the observation that frequency of occurrence of some event f, as a function of its rank i, defined by the frequency, is a power-law function $P_i \sim 1/i^a$ with the exponent a close to unity. Figure 23 shows the word percentage distribution as a function of its frequency rank in our chat corpus.

Figure 23. Word frequency distribution over rank in our chat corpus.

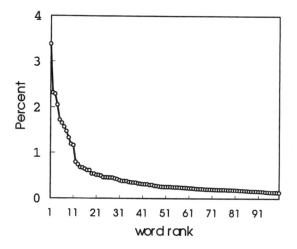

Figure 24. An ATOMIK layout with high degree of efficiency, alphabetical tuning, and word connectivity. The movement efficiency is 41.67 wpm, based on the Mayzner and Tresselt digraph statistics; 41.16 wpm, based on the chat digraph statistics; and 40.78 wpm, based on the news digraph statistics. (Copyright IBM).

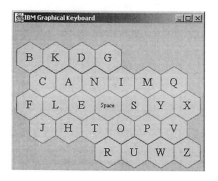

together with the letter keys? Many of the punctuation keys do in fact have higher frequency than some letter keys. For example, the frequency of period (.) is higher than many other alphabetic letters (v, k, j, x, z, etc.; Figures 3 and 4). From the ultimate performance perspective, at least the more frequent punctuation keys should be optimized with the letters key simultaneously.

On the other hand, mixing the punctuation keys with the letter keys will reduce the structural order of the keyboard and increase the search space for the novice users. If all the letter keys are arranged together, the user has to search 26 positions at the most to look for any particular letter key. Otherwise, the user has to search among a greater number of positions, making the keyboard more difficult to learn. It is conceivable that if only a few most frequent keys (e.g., Period and Shift) are mixed with letter keys, the distinct appearance of these keys from the letter keys may become "landmarks" among the letter keys. It has been shown that the presence of visually silent landmarks helps the users to remember the location of the graphical objects around these landmarks (Ark, Dryer, Selker, & Zhai, 1998). However, the cost of mixing a few auxiliary keys with letter keys is the inconsistent organization of keyboard. If all letter keys are together at the core of the keyboard and all auxiliary keys are on the outside, such as in the examples shown in Figures 25 and 26, the more consistent separation may help the novice user to learn about the keyboard. To enhance visual structure, auxiliary keys were given a very different appearance (Figure 26).

According to our digraph table (Appendix Figure A–2), the numeric characters are more frequently connected with each other (and space) than with alphabetic characters. This argues for placing the numeric keys together in a number pad, such as the telephone pad arrangement shown in Figure 25.

Note that the auxiliary keys increase the total number of keys on a keyboard and hence increase the average movement distance, which in turn decreases the total performance of the keyboard. To hold a consistent comparison standard, we always use the letter and Space keys only in the calculation of wpm.

To save space, two auxiliary keys can share one location multiplexed by a Shift key. Figure 25 shows an ATOMIK keyboard implementation with the multiplexing feature. Space saving is particularly necessary on small screens. Figure 26 shows an ATOMIK keyboard implementation on a handheld PDA. Due to the limited screen resolution of the PDA, which causes steps in the rendering of the diagonal edges of hexagons, square-shaped keys were used in this implementation.

5.2. Multiple Space Keys and Varying Key Sizes

Both the OPTI keyboard and the FITALY keyboard used more than one Space key to accommodate the high frequency of space in English writing (Figures 8 and 10). We decided against such an idea for the following reasons. First, multiple Space keys take more space from the real estate available to regular keys, which may reduce the total efficiency of the keyboard. Second, as revealed by our analysis, the performance of a multiple Space key keyboard highly depends on the user's optimal choice of the Space key, which requires planning one key ahead of tapping. Third, the Space key is not the only one with high frequency (Figures 3 and 4).

Related to the multiple Space key issue is the size of the Space key. Should the Space key be given a greater size than other keys? Should other more frequent keys also be given greater size? According to Fitts' law, tapping time is

Figure 25. An ATOMIK keyboard implementation with auxiliary keys (Copyright IBM). Shift key press alternates the functions of the auxiliary keys as well as the letter cases from (a) to (b).

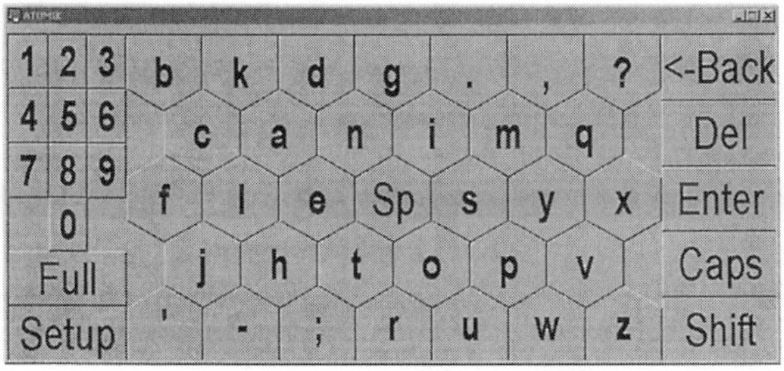

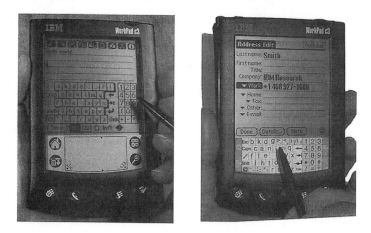

(a)

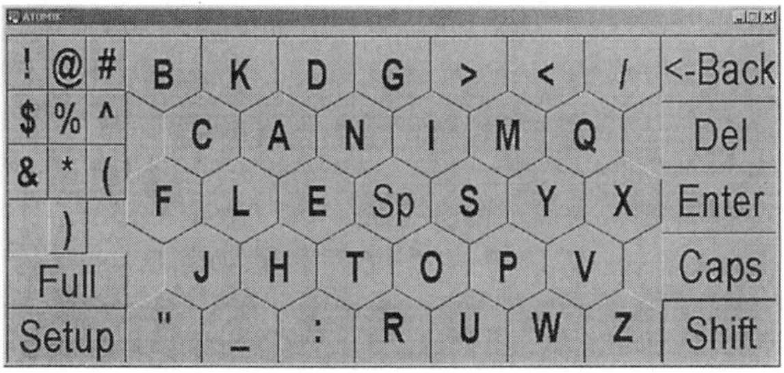

(b)

Figure 26. Two ATOMIK keyboard implementations (Copyright IBM) on a handheld PDA. Some of the keys are multiplexed by the shift or cap key press. The keys are made into square shape in this implementation, which slightly reduces movement efficiency to 39.9 wpm, based on Mayzner and Tresselt digraph statistics.

related to both distance and target size. We have optimized the statistical distance to reduce tapping time. By the same principle, shouldn't we also optimize the relative key sizes so more frequent keys are given a greater share of the real estate?

We have indeed explored the issue of varying key sizes, but we have not come to a positive conclusion. There are at least four issues to resolve. First, the optimization of both size and distance is much more complex. One of the complicating factors is that keys of unequal size cannot be as tightly packed and still be optimized in positional layout. We have made several versions of search algorithms to optimize both the key sizes and the position layout, but so far we have not produced a keyboard that had higher movement efficiency than the Metropolis keyboards with a constant key size.

Second, from Fitts' law point of view, frequently used keys should be given a greater size. However, these frequent keys should also be placed toward the center of the keyboard. Crossing these bigger keys to reach other keys introduces a performance penalty.

Third, there is an asymmetrical effect to size gain and loss in Fitts' law. To reciprocally tap on the two adjacent targets shown in Figure 27, the performance gain of tapping the enlarged right target is less than the loss of tapping the reduced left target. Figure 28 illustrates tapping time from the left to right and vice versa. As we can see, as the asymmetry factor x changes from 0 to positive, the time reduction from the left to the right target does not compensate for the time increase in the reverse direction. The lowest sum of the two is when $x = 0$, assuming equal frequency of entering the left and the right target.

Fourth, even if we found a more efficient layout with varying key sizes, there could be a cost of varying control precision depending on which letter is being typed. The loss of consistency in control precision may be detrimental.

In summary, varying size remains an open problem, although the factors we have considered suggest against it.

5.3. Upper Bounds of Virtual Keyboard Optimization

Has our exploration achieved the maximum efficiency? We have run a large number of simulations, but all reached a similar plateau of movement efficiency, suggesting we are very close, if not at, the maximum efficiency. Figure 29 shows one trace of a Metropolis random walk optimization trial in terms of wpm speed. However, it is theoretically interesting to estimate the lowest upper bound movement efficiency. To that end, we have produced three reference points, based on three physically impossible layout designs. The first was that all keys were co-located in one spot, hence the user only needed to tap on the same spot. With such a hypothetical keyboard, the estimated performance was 95 wpm. The second estimate assumed the next key needed was always next to

Figure 27. Asymmetrical Fitts' targets, with asymmetry coefficient factor *x*.

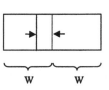

Figure 28. Tapping time from left to right key, vice versa, and the sum of the two as a function of the asymmetry coefficient *x*. The lowest sum occurs when *x* = 0; that is, the two targets are equal in size.

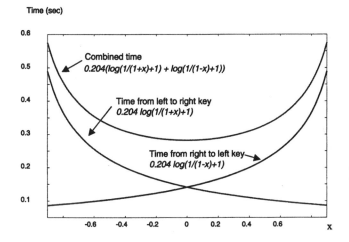

the current key, requiring tapping with Fitts' ID = 1 bit. Performance dropped to 59 wpm with this assumption (at 4.9 bps Fitts' IP). In the third estimate, when calculating Fitts' law performance for any key, the rest of the keys were optimally placed according to their digraph frequency to the current key. The performance of this more realistic but still impossible keyboard was 53 wpm, about 10 wpm faster than our most efficient design. Finding the lowest upper bound of virtual keyboard performance is another open research issue.

6. DISCUSSIONS

6.1. User Interface Design Techniques

Although sharing the same ultimate goal, user interface design and user interface evaluation are traditionally two entirely separate processes involving

Figure 29. A Metropolis random walk trace in terms of wpm speed.

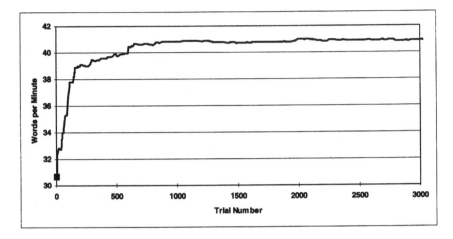

different methodologies. User interface evaluation tends to be measurement based, whereas user interface design tends to be intuition, heuristics, and experience based. The design exploration presented here is a departure from that norm. First, the design process was quantitative and computerized. Second, the design process was integrated to the highest degree with evaluation—every step of the design space search was guided by the evaluation function. Third, the quantitative design process was based on previous evaluation research, particularly the work on Fitts' law. Without Fitts' law, we could still construct an evaluation function simply based on the digraph frequency and travel distance, so we could know the relative superiority of one layout to another based on statistical distance.[6] We would not, however, be able to relate the statistical distance to user performance. With Fitts' law modeled in the evaluation function, we could instantly estimate the eventual average user performance and compare it to known benchmarks.

The optimization methods and software presented here can be used to optimize keyboard layout against any quantifiable objective function and any language corpus. As shown in the design of the ATOMIK layout, we could combine multiple objectives in the same optimization process. However, the relative weights of the difference objectives are design choices that require careful research. In the case of the ATOMIK design, we chose efficiency as the

---

6. Note, however, a layout based on distance optimization may be different from, and less valid than, a layout based on MT optimization due to the nonlinear relation between distance and MT in Fitts' law.

predominant, alphabetical ordering as the secondary, and connectivity index as the third objective.

6.2. Limitations of This Work and Future Research Issues

This study has focused on the optimization of virtual keyboards. We use the term *optimization* in the mathematical sense—searching for the lowest (or highest) value of an objective function with given constraints. The basic objective function we have used is the Fitts-digraph energy defined by Equation 2, which quantifies movement efficiency when using the virtual keyboard with a single stylus. Later, we added alphabetical ordering and word connectivity as additional objectives. The resulting layouts of this optimization process are by no means the "best" in the general sense because there could be many other factors influencing the usability of a virtual keyboard.

Furthermore, movement efficiency in this study was based only on digraphs, without considering how consecutive tapping movements affect each other. It is conceivable that the angle between two consecutive tapping movements has an impact on performance.

We should also reemphasize the assumptions made to calculate the estimated movement efficiency numbers in wpm, which are primarily to serve as indexes for performance comparison. These estimated numbers are not necessarily every user's actual speed at any time. For example, the 41.67 wpm movement efficiency of the ATOMIK layout (Figure 24) was based on the following assumptions:

1. The text entered follows the same digraph distribution of the Mayzner and Tresselt corpus, although we have shown that the Fitts-digraph energy was not very sensitive to the style of text.
2. Every five key strokes, including the Space key, counts for one word.
3. The user tapping performance, as measured by Fitts' law, is at 4.9 bps. If not, the estimated number scales proportionally.
4. The diameter of the inscribed circle of the hexagon keys was taken as the width of target in Fitts' law calculation.
5. The time taken in visual search or other cognitive components in text entering is negligible in comparison to the stylus MT, which is definitely not true for novice users.

A UI design is not complete without user studies. A user study merely to confirm the movement efficiency predictions presented here, however, would not be very informative. First, the very foundation of our calculation, Fitts' law, has already been repeatedly tested. Second, previous user studies have shown that users could indeed master a new layout and eventually reach the performance level of over 40 wpm on the OPTI design (MacKenzie & Zhang, 1999), after twenty 45-min practice sessions.[7] Would an average user be willing to invest 20 hr for 50% performance gain? Twenty hours is probably an unacceptably long period for learning a UI technique, although it is only a fraction of a typical physical Qwerty typist's learning time. We are currently researching training methods that can accelerate the learning curve. Many deeper research questions have to be answered before an effective training method can be found. How do users learn the keyboard layout? Do they learn the paths of words or do they learn the positions of the keys? Our observation to date suggests a combination of both, initially with letter positions and later shifts toward higher level patterns. There is a body of literature on learning typing on a physical keyboard (e.g., Cooper, 1983; Dvorak et al., 1936; Ono & Yamada, 1990), but there is no reason to believe the mechanism involved in tapping on virtual keyboard is the same as 10-finger typing on a physical keyboard. Another class of questions about learning is whether one layout with a particular property is faster to learn than another. We have tried adding landmarks or dividing the keyboard to regions by color coding, but no obvious learning advantage has been gained. There are many fascinating cognitive issues to be investigated in the future.

7. CONCLUSIONS

Motivated by the increasing importance of pervasive computing devices and built on the previous work on virtual keyboards (e.g., Getschow et al., 1986, MacKenzie & Zhang, 1999; Soukoreff & MacKenzie, 1995), this article explored the design of optimized virtual keyboards and made the following contributions. First, the article thoroughly analyzed the movement efficiency of existing virtual keyboards and corrected erroneous estimations of virtual keyboards in the literature. We found the movement efficiency of Qwerty, FITALY, and OPTI keyboards to be in the neighborhood of 28, 36, and 38 wpm, respectively (see Figure 9 for a complete summary). Second, we introduced two quantitative techniques to virtual keyboard design. One technique used physical simulation of digraph springs, producing a Hooke keyboard with 41.6 wpm movement efficiency. The other method used the Metropolis

---

7. A casual user of the layout in Figure 19 "got up to speed" after "about 20 hr of use over 5 days." He wrote a 10-page report and estimated that his speed was "at least twice as fast as Graffiti" (personal e-mail communication, 2001). Although this is totally uncontrolled self-reporting, it corroborates with the learning time reported in MacKenzie and Zhang (1999). We expect the learning speed of the ATOMIK layout incorporating alphabetical tuning and word connectivity to be faster.

random walk algorithm, guided by the Fitts-digraph energy object function. This method produced various Metropolis keyboards around 42.5 wpm movement efficiency, which was more than 50% faster than the Qwerty keyboard. The 42.5 wpm was based on a very conservative assumption of 4.9 bps Fitts' law IP and can be scaled up with the IP value. Third, our design exploration led to a variety of layouts with similar speed performance but different design considerations such as the structure and shape of the overall keyboard. Fourth, in addition to movement efficiency quantified by Fitts-digraph energy, we introduced the concept of alphabetical tuning and produced the ATOMIK layouts, which were demonstrated to be easier for novice users. Fifth, we introduced the concept of word connectivity in the final layout, which may further enhance the usability and acceptability of virtual keyboards. Finally, we illustrated the benefits of quantitative design combined with human performance models over traditional UI design methods based on manual trial and error and heuristics. We demonstrated the importance of quantitative techniques and basic human performance modeling to the field of user interface research.

NOTES

Background. Parts of this article have been previously published in the proceedings of UIST 2000 (Zhai, Hunter, & Smith, 2000), CHI 2000 (Hunter, Zhai, & Smith, 2000), CHI 2001 (Zhai & Smith, 2001), and INTERACT 2001 (Smith & Zhai, 2001).

Acknowledgments. We thank Alison Sue, Teenie Matlock, Jon Graham, and other colleagues at the IBM Almaden Research Center for their input and assistance. We also thank Allison E. Smith for collecting the large Internet relay chat corpus. Comments from the anonymous reviews helped us to improve the article significantly. We particularly thank I. Scott MacKenzie for sharing his spreadsheet model and for numerous fruitful discussions.

Authors' Present Addresses. Shumin Zhai, Department NWE–B2, IBM Almaden Research Center, 650 Harry Road, San Jose, CA 95120. E-mail: zhai@almaden.ibm.com. Michael Hunter. E-mail: michael@hunter.org. Barton A. Smith, Department NEW–B2, IBM Almaden Research Center, 650 Harry Road, San Jose, CA 95120. E-mail: basmith@almaden.ibm.com.

HCI Editorial Record. First manuscript received November 15, 2000. Accepted by Scott Mackenzie. Final manuscript received May 29, 2001. — *Editor*

REFERENCES

Ark, W., Dryer, D. C, Selker, T., & Zhai. S. (1998, March), Landmarks to aid navigation in a graphical user interface. *Proceedings of Workshop on Personalized and Social Navigation in Information Space.* Stockholm, Sweden: ICS.

Beichl, I., & Sullivan, F. (2000). The Metropolis algorithm. *Computing in Science & Engineering, 2*(1), 65–69.

Binder, K., & Heermann, D. W. (1988). *Monte Carlo simulation in statistical physics.* New York: Springer-Verlag.

Cooper, W. E. (Ed.). (1983). *Cognitive aspects of skilled typewriting.* New York: Springer-Verlag.

Dvorak, A., Merrick, N. L., Dealey, W. L., & Ford, G. C. (1936). *Typewriting behavior.* New York: American Book Company.

Fitts, P. M. (1954). The information capacity of the human motor system in controlling the amplitude of movement. *Journal of Experimental Psychology, 47,* 381–391.

Getschow, C. O., Rosen, M. J., & Goodenough-Trepagnier, C. (1986). A systematic approach to design a minimum distance alphabetical keyboard. *Proceedings of RESNA (Rehabilitation Engineering Society of North America) 9th Annual Conference.* Arlington, VA: RESNA.

Hunter, M., Zhai, S., & Smith, B. (2000). Physics-based graphical keyboard design. *Proceedings of the CHI 2000 Conference on Human Factors in Computing Systems.* New York: ACM.

Karat, C.-M., Halverson, C., Horn, D., & Karat, J. (1999). Patterns of entry and correction in large vocabulary continuous speech recognition systems. *Proceedings of the CHI 99 Conference on Human Factors in Computing Systems.* New York: ACM.

Keele, S. W. (1986). Motor control. In K. R. Boff, L. Kaufman, & J. P. Thomas (Eds.), *Handbook of perception and human performance* (pp. 30.1–30.60). New York: Wiley.

Lewis, J. R., Kennedy, P. J., & LaLomia, M. J. (1999). Development of a digram-based typing key layout for single-finger/stylus input. *Proceedings of the Human Factors and Ergonomics Society 43rd Annual Meeting.* Santa Monica, CA: HFES.

Lewis, J. R., LaLomia, M. J., & Kennedy, P. J. (1999). Evaluation of typing key layouts for stylus input. *Proceedings of the Human Factors and Ergonomics Society 43rd Annual Meeting.* Santa Monica, CA: HFES.

MacKenzie, I. S. (1992). Fitts' law as a research and design tool in human computer interaction. *Human–Computer Interaction, 7,* 91–139.

MacKenzie, I. S., Sellen, A., & Buxton, W. (1991). A comparison of input devices in elemental pointing and dragging tasks. *Proceedings of the CHI 91 Conference on Human Factors in Computing Systems.* New York: ACM.

MacKenzie, I. S., & Soukoreff, R. W. (2002). Text entry for mobile computing: Models and methods, theory and practice. *Human–Computer Interaction, 17,* 147–198.

MacKenzie, I. S., & Zhang, S. X. (1999). The design and evaluation of a high-performance soft keyboard. *Proceedings of the CHI 99 Conference on Human Factors in Computing Systems.* New York: ACM.

MacKenzie, I. S., Zhang, S. X., & Soukoreff, R. W. (1999). Text entry using soft keyboards. *Behaviour & Information Technology, 18,* 235–244.

Mayzner, M. S., & Tresselt, M. E. (1965). Tables of single-letter and digram frequency counts for various word-length and letter-position combinations. *Psychonomic Monograph Supplements, 1*(2), 13–32.

McClard, A., & Somers, P. (2000, April). Unleashed: Web tablet integration into the home. *Proceedings of the CHI 2000 Conference on Human Factors in Computing Systems.* New York: ACM.

Metropolis, N., Rosenbluth, A. W., Rosenbluth, M. N., Teller, A. H., & Teller, E. (1953). Equation of state calculations by fast computing machines. *The Journal of Chemical Physics, 21,* 1087–1092.

Norman, D. A., & Fisher, D. (1982). Why alphabetic keyboards are not easy to use: Keyboard layout doesn't much matter. *Human Factors, 24,* 509–519.

Ono, Y., & Yamada, H. (1990). A cognitive type training model whose speed advancement is derived from those of component tasks. *Behavioral Science, 35,* 238–268.

Smith, B. A., & Zhai, S. (2001). Optimised virtual keyboards with and without alphabetical ordering: A novice user study. *Proceedings of Interact 2001—IFIP International Conference on Human–Computer Interaction.* New York: ACM.

Soukoreff, W., & MacKenzie, I. S. (1995). Theoretical upper and lower bounds on typing speeds using a stylus and keyboard. *Behaviour & Information Technology, 14,* 379–379.

Textware Solutions. (1998). *The Fitaly one-finger keyboard.* Retrieved from http://fitaly.com/fitaly/fitaly.htm

Ward, D., Blackwell, A., & MacKay, D. (2000). Dasher: A data entry interface using continuous gesture and language models. *Proceedings of the UIST 2000 Symposium on User Interface Software and Technology.* New York: ACM.

Yamada, H. (1980). A historical study of typewriters and typing methods: From the position of planning Japanese parallels. *Journal of Information Processing, 2,* 175–202.

Zhai, S., Hunter, M., & Smith, B. A. (2000). The Metropolis keyboard: An exploration of quantitative techniques for virtual keyboard design. *Proceedings of the UIST 2000 Symposium on User Interface Software and Technology.* New York: ACM.

Zhai, S., & Smith, B. A. (2001). Alphabetically biased virtual keyboards are easier to use: Layout does matter. *Proceedings of the CHI 2001 Conference on Human Factors in Computing Systems.* New York: ACM.

Zhang, S. X. (1998). *A high performance soft keyboard for mobile systems.* Unpublished master's thesis, The University of Guelph, Ontario, Canada.

APPENDIX

Digraph Tables Compiled From Text Available on the Internet

Figure A-1. Digraph table of the 26-letter alphabet and the 10 most frequent nonalphabet letters in the chat room corpus.

	space	!	"	'	,	-	.	1	:	?	a	b	c	d	e	f	g	h
space	4009	65	1894	215	107	648	675	1359	952	185	21745	9578	9261	7650	3973	6532	6755	10090
!	487	1377	263	4	0	0	7	8	3	65	0	29	0	12	0	0	1	1
"	884	0	2	1	31	1	259	2	0	28	150	78	62	81	40	53	78	141
'	233	0	1	2	7	1	22	0	0	12	41	21	21	217	21	13	13	9
,	9888	0	16	3	21	12	9	384	0	1	9	3	3	3	2	0	1	1
-	882	0	1	7	10	526	18	95	17	1	51	60	92	65	35	42	42	34
.	5105	5	777	12	14	4	13069	19	41	47	304	209	153	97	75	54	89	237
1	708	13	148	1	145	34	40	141	175	3	10	2	9	3	3	1	4	55
:	2331	2	7	0	0	108	0	221	256	1	4	1	2	17	0	0	1	2
?	1019	105	325	5	2	0	52	0	0	1845	1	0	0	0	0	4	0	2
a	7498	86	27	53	259	79	304	6	128	112	465	1809	2574	2828	95	402	1543	1653
b	245	16	8	6	33	10	50	0	14	16	1810	290	34	115	5033	8	9	10
c	777	8	6	18	97	12	199	0	53	49	4074	15	446	46	3020	2	9	5429
d	14705	114	94	141	696	58	1065	5	160	262	1610	34	108	305	4356	48	213	51
e	34370	375	140	514	1271	175	2137	44	272	761	6619	426	2027	5349	3716	990	620	1038
f	5784	20	9	10	73	52	156	3	42	52	1597	0	6	20	1523	1122	89	2
g	5761	57	27	23	249	24	432	2	32	117	1275	12	5	60	2665	13	294	1875
h	4624	101	29	29	526	45	549	9	60	152	12788	52	33	37	21947	22	9	414
i	6295	32	15	1027	99	64	123	0	43	50	1759	687	4239	2615	3136	2381	2018	35
j	51	3	0	2	4	4	13	0	19	0	486	3	11	2	1098	4	5	7
k	2491	72	13	9	178	19	241	8	33	91	701	17	6	11	3063	15	14	381
l	5774	90	38	40	488	95	585	18	113	130	3227	230	165	2253	6082	542	144	17
m	4209	47	20	24	358	24	549	14	71	148	4265	412	74	60	7113	28	13	16
n	12907	146	57	2027	726	149	927	6	169	272	3056	77	1760	7938	5569	265	7641	43
o	10884	133	12	49	394	72	571	13	63	171	282	794	1164	3161	863	4755	630	983
p	1205	33	12	5	95	22	189	7	93	49	1441	12	20	6	3161	12	8	708
q	10	0	0	0	3	2	2	0	2	0	8	2	0	0	1	0	0	0
r	8898	85	51	77	444	68	603	24	105	135	3884	310	750	1500	13277	210	727	100
s	22762	204	137	101	1381	68	1820	23	221	497	3168	111	934	57	6075	58	647	2816
t	23464	159	96	1072	908	110	1415	7	153	592	3453	60	457	45	6273	48	61	25793
u	4320	37	14	332	126	39	165	0	64	84	686	391	1247	713	1016	411	1001	174
v	310	7	0	2	3	2	20	1	2	4	474	6	7	2	7300	1	5	5
w	2298	33	14	25	155	9	304	0	5	71	4759	94	3	21	3235	18	7	4332
x	251	5	5	4	31	21	39	0	11	34	234	0	114	6	125	1	2	5
y	11413	149	52	174	856	66	865	17	173	325	514	272	68	25	1784	101	118	35
z	191	15	0	4	49	13	30	0	58	14	94	0	0	3	227	0	1	2

i	j	k	l	m	n	o	p	q	r	s	t	u	v	w	x	y	z
17781	2193	2323	5576	9862	5798	10647	6528	473	4274	12637	28897	2468	1107	14233	116	6056	75
2	0	4	17	0	3	1	10	1	7	9	1	0	0	2	0	1	0
267	23	18	31	81	108	86	56	4	48	153	168	35	14	214	5	105	1
14	0	2	239	656	29	12	10	0	375	1872	1915	8	212	8	0	6	0
7	0	1	2	22	0	1	5	0	3	0	9	11	2	4	1	2	0
24	43	19	32	70	31	51	47	0	37	82	60	24	20	46	2	45	4
513	43	18	66	155	144	177	52	3	58	175	308	66	21	222	8	173	3
1	6	2	0	2	2	13	1	0	2	19	2	56	0	8	0	0	0
5	1	0	2	0	2	16	186	0	0	4	6	0	0	3	0	3	0
1	0	0	0	12	1	0	0	0	3	2	0	0	1	2	0	0	0
2545	100	1584	7926	2721	17032	71	1150	7	8083	6395	12701	1425	2409	539	87	2854	399
1338	145	4	1831	7	30	2337	6	0	895	257	99	1994	71	24	0	696	10
1316	3	2494	829	8	12	3576	13	6	1083	481	1771	688	3	0	0	186	26
2814	14	3	228	130	520	4263	6	5	749	1038	27	758	50	27	1	582	15
1683	49	342	5488	2220	7563	1049	958	96	13449	8729	3606	181	2544	1036	1313	2232	106
1546	3	15	405	15	9	2872	1	0	1444	64	375	1046	0	2	9	70	0
2153	4	4	546	59	292	3949	0	1	1184	567	43	1108	4	8	0	181	1
7138	6	26	57	234	342	4669	12	3	1041	117	1359	1137	13	48	8	775	0
410	16	1141	3110	2719	16266	3034	633	93	2734	9766	9615	86	1543	27	106	42	203
38	16	8	0	0	5	596	3	0	18	9	11	1072	169	18	0	0	0
2009	169	21	86	13	1171	300	10	0	80	944	63	72	5	50	0	203	0
5395	1	543	7169	143	27	4721	214	0	150	1510	429	716	182	175	6	3350	92
1812	1	10	61	1022	219	2506	876	0	31	1145	54	1037	0	8	1	1208	1
2349	97	1359	616	75	1453	6352	51	19	108	2630	5478	448	180	37	28	2539	21
974	95	1122	3200	4208	11489	4163	1896	6	8040	2002	4866	11567	1130	3698	115	327	80
1724	2	13	2029	46	2	1902	935	0	1911	567	598	581	1	0	1	153	0
0	0	0	1	1	1	1	0	0	0	1	0	768	1	2	0	0	0
4628	11	646	1229	511	909	4490	234	1	901	2702	1991	1387	267	133	10	1974	9
2744	11	364	595	634	616	3901	1177	30	52	2298	8368	2183	17	312	5	356	10
6133	13	12	677	107	92	8430	87	2	2334	2391	1605	1493	49	326	7	1268	38
591	19	197	2667	946	2631	137	1182	2	3885	5087	4058	98	11	10	32	492	69
1583	0	0	8	3	9	489	0	0	7	16	8	7	2	0	0	10	0
3350	5	9	138	7	570	2410	5	0	382	365	34	73	1	242	0	19	1
330	0	1	6	0	0	17	250	0	2	7	197	42	0	3	88	91	0
406	1	82	66	147	61	6542	127	1	82	935	401	182	2	130	1	131	19
124	0	34	6	2	5	55	0	0	1	2	3	27	0	1	0	96	132

Figure A–2. Digraph table of the 26-letter alphabet and the 10 most frequent nonalphabet letters in the news corpus.

	space	"	'	'	-	.	0	1	2	'	a	b	c	d	e	f	g	h
space	20	62	3	2	56	19	2	149	58	38	2069	1034	1104	539	454	817	385	762
"	88	0	0	0	0	0	0	0	0	0	5	6	6	1	1	1	3	1
'	101	0	85	0	0	0	0	1	0	1	0	0	0	3	0	0	0	0
,	1103	45	47	0	0	0	11	2	1	0	0	0	0	0	0	0	0	0
-	63	0	0	0	43	0	2	13	7	0	7	15	8	10	4	2	10	3
.	1050	31	33	13	1	21	2	7	7	0	1	1	9	0	0	1	1	3
0	69	0	6	7	8	11	65	2	1	0	0	0	0	0	0	0	0	0
1	25	0	0	13	5	8	23	6	8	0	0	0	0	0	0	0	0	0
2	23	0	0	7	5	8	20	4	6	0	0	2	0	0	0	0	0	0
'	0	0	0	0	0	0	0	0	0	81	5	3	1	1	0	2	1	4
a	622	0	18	31	2	15	0	0	0	0	1	126	288	369	6	78	171	12
b	24	0	1	6	0	7	0	0	2	0	185	21	1	2	417	0	0	0
c	76	1	1	2	0	16	0	0	0	0	402	1	63	1	500	0	0	410
d	1917	0	8	99	13	119	0	0	0	0	208	1	2	25	619	3	12	3
e	3354	2	27	171	18	129	0	0	0	0	606	57	347	954	295	179	87	19
f	629	0	0	6	17	11	0	0	0	0	106	1	0	0	194	156	2	0
g	577	1	2	30	1	25	0	0	0	0	186	0	0	1	260	0	27	198
h	516	1	10	20	9	21	0	0	0	0	658	7	0	10	2039	0	15	0
i	114	0	11	10	1	12	0	0	0	0	194	55	518	387	239	115	231	0
j	0	0	0	0	0	3	0	0	0	0	19	0	0	0	29	0	0	0
k	170	0	2	24	2	19	0	0	0	0	68	1	0	3	272	0	0	3
l	569	1	10	47	13	24	0	0	0	0	496	6	2	231	610	33	2	1
m	220	0	7	18	3	28	0	0	0	0	495	79	4	2	643	3	0	2
n	1579	2	64	112	15	89	0	0	0	0	348	4	231	896	578	53	803	3
o	758	0	5	33	13	18	0	0	0	0	71	53	163	141	18	641	33	41
p	121	0	1	13	1	31	0	0	0	0	240	0	1	1	355	0	0	55
q	4	0	1	2	0	1	0	0	0	0	0	0	0	0	0	0	0	0
r	990	1	14	94	23	123	0	0	0	0	444	16	139	116	1401	22	97	3
s	2263	0	27	205	4	252	0	0	0	0	354	4	97	48	683	3	1	354
t	1780	0	38	94	22	85	0	0	0	0	376	3	34	2	952	3	1	2353
u	39	0	8	5	0	35	0	0	0	0	78	84	105	53	112	16	68	0
v	16	0	1	2	0	6	0	0	0	0	58	0	0	0	554	0	0	0
w	147	0	3	6	3	22	0	0	0	0	318	1	1	7	282	1	0	217
x	28	0	1	2	1	2	0	0	0	0	16	0	11	0	10	0	1	1
y	912	0	26	85	2	73	0	0	1	0	44	5	7	3	111	0	1	2
z	7	0	1	4	0	5	0	0	0	0	35	0	0	0	48	0	1	1

i	j	k	l	m	n	o	p	q	r	s	t	u	v	w	x	y	z
1220	124	90	527	804	396	117	944	30	636	1468	2810	243	109	1118	0	153	6
15	0	0	3	3	3	4	4	0	2	2	14	0	0	14	0	2	0
0	0	0	4	2	4	0	0	0	8	214	30	0	6	1	0	0	0
0	0	0	0	0	0	0	0	0	0	0	0	0	0	0	0	0	0
2	1	1	7	9	2	18	14	2	10	16	13	4	2	3	0	8	0
9	1	1	4	8	9	3	11	0	1	19	0	0	1	1	0	1	0
0	0	0	0	0	0	0	0	0	0	0	3	0	0	0	0	0	0
0	0	1	0	0	0	0	0	0	0	0	0	0	0	0	0	0	0
0	0	2	0	0	0	0	0	0	0	0	0	0	0	0	0	0	0
19	0	0	2	3	0	1	2	0	1	6	24	0	0	6	0	3	0
424	19	136	727	281	1433	2	123	10	760	767	1114	99	146	51	6	317	13
114	1	0	182	13	0	160	0	0	138	17	5	188	1	0	0	159	0
183	0	172	149	1	0	697	0	3	111	12	244	83	0	0	0	29	1
245	0	0	24	28	25	145	0	1	77	83	3	85	25	10	0	33	0
110	13	34	386	219	1076	58	140	26	1553	989	266	27	177	119	129	96	12
255	0	0	67	0	0	421	0	0	131	5	97	51	0	0	0	4	0
120	0	3	36	1	56	131	0	0	166	29	23	79	0	0	0	11	0
512	0	0	12	6	36	323	0	0	60	16	131	45	0	7	0	16	0
2	3	45	357	186	2025	474	47	1	259	737	803	12	175	1	18	0	68
20	0	0	0	0	0	82	0	0	0	0	0	46	0	0	0	0	0
101	0	0	7	1	20	35	3	0	9	62	0	2	0	0	0	1	0
508	0	15	419	15	2	269	20	0	6	151	44	71	11	5	0	274	1
269	4	0	1	78	3	271	174	0	66	41	0	54	0	1	0	39	0
302	14	72	49	53	71	282	14	4	5	348	833	70	64	1	0	95	12
99	7	65	199	452	1296	169	207	0	1019	219	344	661	171	264	3	20	4
69	0	1	206	6	0	343	108	0	382	47	49	119	0	0	0	6	0
0	0	0	0	0	0	0	0	0	0	0	0	80	0	0	0	0	0
553	1	121	88	105	152	576	39	1	72	375	248	104	47	5	0	127	0
425	3	36	46	71	14	242	132	9	1	324	850	238	0	13	0	40	1
809	0	0	65	13	10	908	10	0	247	339	138	158	1	56	0	93	6
66	1	5	251	96	344	2	118	1	343	335	355	0	5	0	4	6	0
237	0	0	2	0	0	63	0	0	0	0	1	0	0	0	0	1	0
318	0	0	16	4	70	205	2	0	18	41	9	0	0	10	0	4	0
8	0	0	0	0	0	1	46	0	0	0	30	2	0	0	0	0	0
29	0	3	8	13	9	117	7	0	0	68	5	2	0	0	0	0	0
15	0	0	0	0	0	2	0	0	0	0	0	3	0	0	0	2	2

HUMAN-COMPUTER INTERACTION, 2002, Volume 17, pp. 271–309

Empirical Bi-Action Tables: A Tool for the Evaluation and Optimization of Text-Input Systems. Application I: Stylus Keyboards

Dominic Hughes, James Warren, and Orkut Buyukkokten
Stanford University

ABSTRACT

We introduce a technique that, given any text input system A and novice user u, will predict the peak expert input speed of u on A, avoiding the costly process of actually training u to expert level. Here, *peak* refers to periods of ideal performance, free from hesitation or concentration lapse, and *expert* refers to asymptotic competence (e.g., touch typing, in the case of a two-handed keyboard). The technique is intended as a feedback mechanism in the interface development cycle between abstract mathematical modeling at the start (Fitts' law, Hick's law, etc.) and full empirical testing at the end.

Dominic Hughes is a computer scientist specializing in logic and mathematical foundations of computation; he is a Research Associate in the Computer Science Department of Stanford University. **James Warren** is a computer scientist with interests in machine learning and optimization; he is a PhD student in the Scientific Computing and Computational Mathematics Program of Stanford University. **Orkut Buyukkokten** is a computer scientist with interests in databases and human–computer interaction; he is a PhD student in the Computer Science Department of Stanford University.

CONTENTS

The utility of the technique in iterative design is contingent on what we call the monotonicity principle: For each user u, if our prediction of peak expert input speed for u is higher on system A than on system B, continuous text input by u after training to expert level will be faster on A than on B. Here, *continuous* refers to actual real-world use, subject to errors, physical fatigue, lapses of concentration, and so forth. We discuss the circumstances under which monotonicity is valid.

The technique is parametric in the character map—that is, in the map from actions (keystrokes, gestures, chords, etc.) to characters. Therefore, standard heuristic algorithms can be employed to search for optimal character maps (e.g., keyboard layouts). We illustrate the use of our technique for evaluation and optimization in the context of stylus keyboards, first benchmarking a number of stylus keyboards relative to a simple alphabetic layout and then implementing an ant algorithm to obtain a machine-optimized layout.

1. INTRODUCTION

As with any design process, the design of text input systems requires a feedback mechanism to iterate to better solutions. Full empirical evaluation of expert performance on a new text input system is costly, due to the vast number of hours required to train test participants to expert level. Furthermore, evaluation is highly sensitive: Even a minor modification in the system (e.g., switching the position of a key on a keyboard layout, changing a chord on a chording keyboard, or modifying a gesture of a glove input language) forces a repeat of all experiments because participants must be retrained to expert level on the modified system. Such high cost and sensitivity render the iterative design of text input systems impractical; interface design becomes more of an art than a science.

In certain cases, rules, models, or equations (e.g., Fitts' law, Hick's law, and the power law; Card, Moran, & Newell, 1983) can be used to generate a feedback loop in the early stages of the design process. However, these techniques are not without drawbacks. First, by the very nature of abstraction, there can be problems of fidelity and resolution (Noel & McDonald, 1989). Second, there is the problem of lack of generality: Some systems may be beyond the scope of laws. For example, how does one model intricate gestures with a glove or stylus, or the complex parallelism and interference between fingers at a two-handed keyboard?

We introduce a technique intended as a tool in the interface design cycle between abstract mathematical modeling at the start (Fitts' law, etc.) and full empirical user testing at the end. Given any text input system A and novice user u, the technique predicts the peak expert input speed of u on A, avoiding the costly process of actually training u up to expert level. Here, *peak* refers to periods of ideal performance, free from hesitation or concentration lapse, and *expert* refers to asymptotic competence (e.g., touch typing in the context of a two-handed keyboard).

Our conceptual starting point is a strict separation of the text input system into two parts:

- *Physical aspect:* The *actions* performable on the device—for example, a keystroke of a two-handed keyboard, the articulation of a gesture with a stylus or glove, or the depression of a chord on a two-handed keyboard.
- *Logical aspect:* The *character map*, specifying the interpretation of each action as a character—for example, "striking the top left key" maps to Q, "a vertical down-stroke with the index finger of the glove" maps to I, "chording the two outermost keys" maps to Y.

For each (potentially novice) user u and text input system A we capture the pure physical aspect of interaction between u and A, in total isolation from the logical aspect, as an empirical bi-action table E. For each pair of actions i and j (e.g., keystrokes on a two-handed keyboard, gestures with a stylus or glove, etc.), the table entry E_{ij} is the aggregate result of experiments recording the time to complete j having just completed i, in the absence of any character map. Then, given an arbitrary character map, we obtain a prediction of peak expert input speed by using a table of character bi-gram frequencies (see Figure 1).

It is crucial that the experiments are completely independent of the logical aspect (i.e., are conducted in the absence of a character map). For example, in the context of glove gestures we might ask a user to form a fist then flatten the hand—without reference to any particular interpretation of these actions as characters. In the case of stylus keyboards, we might place users at a completely blank grid of keys and ask them to tap the top left key followed by the bottom right key. The reasoning behind our choice to conduct pure physical experimental trials in the absence of a character map is as follows:

1. The experimental data capture physical coordination representative of expert-level production of text during peak concentration. Pairs of actions are executed fluently, as they would be by an expert user who is completely familiar with an ambient character map—without actually having to train the user to expert level on a character map.
2. Once the empirical bi-action table is obtained, we can immediately predict peak expert input speed for the device under any character map.
3. Because predictions are obtained immediately for any character map, we can employ standard heuristic algorithms to search for an optimal character map (e.g., keyboard layout).

Our prediction of peak expert input speed is not intended to be an estimate of real continuous performance. The latter is subject to errors, physical fatigue, and lapses in concentration, factors that are highly unpredictable, varying not only between users but also on a session-by-session basis for a single user. However, in the absence of realistic predictions of continuous input speeds, our technique is nonetheless a useful feedback mechanism for the iterative design of text-input systems if one accepts the validity of the *monotonicity[1] principle*, which we present in a strong and a weak form:

> *Strong Monotonicity Principle*: For any user u and any text input systems A and B, if our prediction of peak expert input speed for u is higher on A than on B, continuous text input by u after training to expert level will be faster on A than on B.

---

1. The terminology derives from mathematics. A function $f: X \rightarrow Y$ between ordered sets X and Y is monotonic if and only if $\forall\ x,\ x' \in X.\ [x \le x' \Rightarrow fx \le fx']$.

Figure 1. Predicting peak expert text input rates.

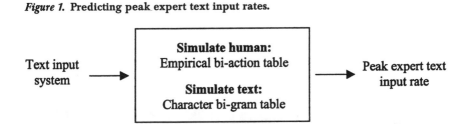

Weak Monotonicity Principle: For any user *u* and text input systems *A* and *B* that differ only in character map (and hence have the same empirical bi-action table), if our prediction of peak expert input speed for *u* is higher on system *A* than on system *B*, continuous text input by *u* after training to expert level will be faster on *A* than on *B*.

In Section 2 we outline why the strong form does not hold in general, particularly when *A* and *B* are physically very dissimilar. Therefore, when using our predictions of peak expert input speed to benchmark one text input system *A* against a very different system *B* (e.g., Morse code against handwriting recognition), careful consideration is required on the part of the researcher before the results can be deemed meaningful.

Also in Section 2, we argue that the weak monotonicity principle holds. Consequently, it is valid to use the empirical bi-action technique to evaluate and compare alternative character maps for the same device and to search for optimal character maps using heuristic algorithms. We demonstrate this approach in Section 3 in the context of stylus keyboard layouts. We benchmark a number of stylus keyboards relative to a simple alphabetic layout, then implement an ant algorithm to obtain a machine-optimized layout. In particular, we validate our technique by successfully correlating predicted peak expert input rates with previous results on two stylus keyboards: the OPTI (MacKenzie & Zhang, 1999) and the FITALY (Textware™ Solutions, 2000). We summarize this illustrative example later.

In Section 4, we discuss the strengths and weaknesses of our technique relative to full empirical testing and abstract mathematical modeling (Fitts' law, etc.). Having illustrated the approach in the context of stylus keyboards, we note that empirical bi-action tables could be used in the analysis and design of a wide variety of text input systems, such as two-handed keyboards, chording keyboards, cell phones, glove gesture input, and forms of stylus input including Graffiti®, Quikwriting (Perlin, 1998), and Unistrokes (Goldberg & Richardson, 1993).

In Section 3, we illustrate our technique in the context of stylus keyboard design. A stylus keyboard is a graphical keyboard displayed on a touch screen, on which users type by tapping with a stylus (pen). An example is the pop-up

Qwerty available on many personal digital assistants (PDAs). Other methods of text entry with a stylus include handwriting recognition and gesture recognition.

It is a point of contention as to whether users of mobile devices spend enough time entering text to be willing to expend the effort to acquire expertise with a faster layout than the familiar Qwerty. We justify the search for faster layouts as follows: (a) There is a market, as witnessed by sales of a commercial layout called the FITALY (Textware Solutions, 2000); (b) although a large volume of text will rarely be entered in a single session, high numbers of short messages are likely; and (c) only *after* researchers have explored the space of optimized keyboards and have understood what can be gained by switching from Qwerty can we conclude that a majority of users would continue using the Qwerty. We do not dwell on these issues in this article because we are using stylus keyboard design as a domain to illustrate a more general approach.

Both full empirical testing and abstract mathematical modeling have been used in stylus keyboard design (Hunter, Zhai, & Smith, 2000; Lewis, LaLomia, & Kennedy, 1999; MacKenzie, Nonnecke, McQueen, Riddersma, & Meltz, 1994; MacKenzie & Zhang, 1999; MacKenzie, Zhang, & Soukoreff, 1999; Soukoreff & MacKenzie, 1995). Abstract approaches use an equational characterization of human motion (known as Fitts' law; Fitts, 1954) to simulate user input at keyboards and hence obtain estimates of text entry rate. Then, with an evaluation function at hand, one can apply off-the-shelf techniques to find optimized keyboard layouts. For example, Hunter et al. (2000) employed dynamic simulation and the Metropolis method (see also Zhai, Hunter, & Smith, 2000). A drawback of the pure analytical approach is that there is experimental evidence (Section 2.2 of MacKenzie, 1991) that small-scale hand motions are not accurately characterized by the law.

Due to the inherently abstract nature of Fitts' law, previous works have only considered the distance between two keys as a predictor of the duration of the motion between these keys. We show that this duration depends also on the first key position and on the relative position of the second key. These dependencies, as well as any other more subtle dependencies that could be impossible to model, are automatically taken into account by our empirical bi-action table, supporting its use in the intermediate design phase between an initial use of laws and the final full user testing.

In the context of stylus keyboards, an action is a tap of the stylus, so in Section 3 we refer to bi-taps instead of bi-actions. In Section 3.9 we describe the implementation of an ant algorithm to find an optimized stylus keyboard layout, and in Section 3.11 we benchmark a number of keyboards against a naive alphabetical layout ABC. The peak input rate of the layout produced by the ant algorithm was 15.65% faster than the ABC, the FITALY was 13.35% faster, the OPTI was 11.63% faster, and a variant of the ABC with a center Space key was 3.95% faster. Until a larger corpus of bi-tap data has been amassed, these figures should not be considered final because only five participants were tested in the original construction of the bi-tap table. Furthermore, the output of the ant algorithm had the advantage of the coincidence of training data and test data.

2. EMPIRICAL BI-ACTION TABLES

Given a text input system S, we perform experiments to capture the physical aspect of S in an empirical bi-action table E: For each pair of actions i and j (e.g., keystrokes on a two-handed keyboard, gestures with a stylus or glove, etc.), the entry E_{ij} is the aggregate result of experiments recording the time to complete j having just completed i. For the reasons outlined in Section 1, it is crucial that the experiments be conducted in complete isolation from the logical aspect (i.e., completely independently of any character map). An illustration of how to perform such experiments is given in detail in Section 3, in the context of stylus keyboards.

Given a character map K (i.e., an assignment of actions to characters), a table of bi-gram probabilities P and an empirical bi-action table E,

Character map K: character a $\rightarrow$ action $K(\alpha)$
Bi-grams P: (character α, character β) $\rightarrow$ probability $P(\alpha, \beta)$
Bi-actions E: (action i, action j) $\rightarrow$ duration $E(i, j) = E_{ij}$

the peak expert text input rate $R(K, P, E)$, in characters per unit time, is given by:

$$R(K,P,E) = \frac{1}{\sum_{\alpha,\beta} P(\alpha,\beta)E(K(\alpha),K(\beta))} \tag{1}$$

where α and β range over the character set.

This equation can be decomposed and understood as follows:

(α, β) *bi-gram execution time* $M_{P,E}(\alpha,\beta) = E(K(\alpha),K(\beta))$

Mean bi-gram execution time $\bar{M}(K,P,E) = \sum_{\alpha,\beta} P(\alpha,\beta)M_{P,E}(\alpha,\beta)$

Peak expert text-input rate $R(K,P,E) = \frac{1}{\bar{M}(K,P,E)}$

$\bar{M}(K, P, E)$ is the mean time taken to input an ordered pair of characters (bi-gram) under the character map K, with text represented by the bi-gram probabilities P and motion modeled by the empirical bi-action table E. $M_{P,E}(\alpha, \beta)$ is the time taken to input the ordered pair of characters (α, β) under the character map K, with motion modeled by the empirical bi-action table E. Note that this depends only on the character map K and empirical bi-action table E.

From this decomposition we observe that our Equation 1 for text-input rate $R(K, P, E)$ is similar to Equation 5 of MacKenzie et al. (1999) with (α, β) bi-gram movement time set to $E(K(\alpha), K(\beta))$.

The peak expert text input rate defined earlier is not intended to be an estimate of real continuous performance. Actual text input, being subject to factors such as physical fatigue and lapses in concentration, will only reach peak rate in short bursts. The peak expert text input rate is thus a useful mechanism for the iterative design of text input systems only if one accepts the validity of the *monotonicity principle*, which we present in a strong and a weak form:

> *Strong Monotonicity Principle*: For any user u and any text input systems A and B, if our prediction of peak expert text input rate for u is higher on A than on B, continuous text input by u after training to expert level will be faster on A than on B.
>
> *Weak Monotonicity Principle*: For any user u and text input systems A and B that differ only in character map (and hence have the same empirical bi-action table), if our prediction of peak expert text input rate for u is higher on system A than on system B, continuous text input by u after training to expert level will be faster on A than on B.

The strong form does not hold in general, particularly when A and B are physically very dissimilar. The sources of fatigue may be very different for different devices, as may be concentration levels required during use. Consider, for example, the case of A, an optimized gesture language such as Unistrokes (Goldberg & Richardson, 1993), and B, a hunt-and-tap stylus keyboard. Due to the continuous visual scanning required during input at the stylus keyboard, actual sustained input on B may be more prone to fatigue or errors, so a slightly higher prediction of peak expert input speed for B than for A may not accurately represent better real-world sustained performance on B than on A. Therefore when using our predictions of peak expert input speed to benchmark one text input system A against a very different system B, careful consideration is required on the part of the researcher before the results can be deemed meaningful.

However, when A and B use the same device and set of actions and differ only in character map, factors such as concentration laps and fatigue will be similar. Therefore, continuous text entry rate will be a similar dilution of peak text entry rate in each case, and the Weak Monotonicity Principle holds. Consequently, it is valid to use the empirical bi-action technique to evaluate and compare alternative character maps for the same device and to search for optimal character maps using heuristic algorithms. We demonstrate this approach in Section 3 in the context of stylus keyboard layouts.

3. ILLUSTRATIVE EXAMPLE: STYLUS KEYBOARDS

We illustrate the use of an empirical bi-action table in the context of stylus keyboard evaluation and optimization. We describe in detail the experiment to generate the empirical bi-action table, benchmark various layouts against a simple alphabetical layout, and then we implement an ant algorithm to search for an optimal layout. We validate our technique by correlating predicted peak expert input speeds with previous results on the OPTI and FITALY layouts. Because actions are taps with a stylus, in this section we refer to bi-actions as bi-taps.

3.1. Method

Participants. There were five participants, each right-handed. All were students, familiar with desktop computing. Four were men. The youngest was 20, the oldest was 32, and the mean age was 26.3. Of the 5, only 1 had previous experience with stylus text input. All were paid for their participation in the study.

Apparatus. For the experiments we used several different PDAs: Palm™ III, Palm IIIx, Palm VII (Palm, Inc.), and Visor (Handspring™). The software was written in C++ using CodeWarrior™ for Palm OS® (Metrowerks). The time measurements were gathered using the device clock, in terms of *time ticks*. The operating system has a GetTimeTicks() function that gives the time elapsed in milliseconds. This function is called each time the user taps on the screen, and the time elapsed between two pen taps is measured by taking the difference between two consecutive values. The data were saved in a Palm OS database and later downloaded to a desktop computer for analysis.

Procedure. Each participant underwent five separate tests. The duration of a test was approximately 30 min. Participants worked on a 5-row, 6-column blank keyboard on a PDA (Figure 2). The dimensions of the grid were as follows: width $= 3.15$ cm, height $= 2.65$ cm, square key width/height $= 5.25$ mm.

Define a *bi-tap* to be any ordered pair of keys (k_1, k_2). A test consisted of presenting a participant with all 900 possible bi-taps in random order. Each bi-tap was presented by labeling two of the blank keys, 1 and 2. The participant tried as

Figure 2. Screen shots from the bi-tap experiment. The screen shots are in order of bi-taps presented to a user during the test (observe the count in the top of the display). The lower right shot shows the irritating 3-sec screen-lock incurred by committing an error.

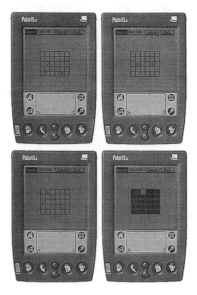

quickly as possible to tap key 1 followed by key 2. We recorded the time of each such bi-tap (i.e., the time interval between tapping key 1 and tapping key 2).

Participants were aware of the fact that their time to find and hit key 1 is not being recorded. Furthermore, they were instructed to absorb the positions of both keys before undertaking any physical motion, so that the recorded interval does not include scanning time for key 2. Should a participant be interrupted or distracted after striking key 1, they are instructed to the cancel button to be presented with the same bi-tap a second time. When an incorrect key was tapped the application emitted an ugly BEEP, and the screen locked for 3 sec. The irritation ensured an extremely low error rate.

An important aspect of the test was that all 900 bi-grams were carried out in succession rather than in isolation. This was to simulate the fact that during real typing, participants typically adopt a natural rest position with their wrist on the side of the PDA. For consistency, all participants were asked to work the PDA in hand rather than supported flat on a desk.

3.2. Results

We adopt the chess naming convention for keys as depicted in Figure 3, row index A to F from left to right, column index 1 to 5 from bottom to top.

Figure 3. Naming convention for keys.

A5	B5	C5	D5	E5	F5
A4	B4	C4	D4	E4	F4
A3	B3	C3	D3	E3	F3
A2	B2	C2	D2	E2	F2
A1	B1	C1	D1	E1	F1

The results of the experiment, aggregated across the 25 tests of the 5 participants, are shown in Figure 4, which we refer to as the empirical bi-tap table. The entry at row k_1 and column k_2 represents the time taken by a generic user between tapping key k_1 and key k_2. The largest entry in the table is .330 sec, for the long diagonal bi-tap A1 to F5. The smallest entry is .147, for the double-tap of F3.

Note that we use $\bar{E}$ to denote the empirical bi-tap table of Figure 4. As detailed in the next section, the overline is to remind us that $\bar{E}$ is the mean of five bi-tap tables, one per test participant. We write $\bar{E}(k_1, k_2)$ for the entry at row k_1 and column k_2.

3.3. Calculating the Entries of the Empirical Bi-Tap Table

Having undergone the experiment five times, each user generated five data points per bi-tap; hence, we have a total of 25 data points per bi-tap. During the testing procedure we occasionally observed lapses of concentration by a user in the middle of a bi-tap. Such instances result in anomalous data points that are not in agreement with our objective of capturing the purely physical "minimum transition time" between a pair of taps.

Outliers were discarded uniformly with the following procedure: For each user, and for each bi-tap, discard the data points that are more than twice the duration of the minimum bi-tap time in the user's quintuple of recorded data points. From a total set of 21,750 data points for bi-taps consisting of distinct keys, this procedure discards 877 points (i.e., an outlier cutoff of 4%). Note that we remove outliers on a per user basis to allow for the fact that some users are uniformly faster and more coordinated than others.

Figure 4. Empirical bi-tap table $\bar{E}$. The value $\bar{E}(k_1, k_2)$ at row k_1 and column k_2 is the time taken to tap key k_1 followed by key k_2 in seconds. (See Figure 3 for the naming convention for the keys on the 5 × 6 grid.) The value of $\bar{E}(k_1, k_2)$ is the mean of the corresponding values in the bi-tap tables of the 5 participants. Section 3.3 explains how the value is computed from the 25 experimental data points for (k_1, k_2).

1st/2nd	A1	B1	C1	D1	E1	F1	A2	B2	C2	D2	E2	F2	A3	B3	C3
A1	0.157	0.157	0.188	0.225	0.254	0.281	0.165	0.175	0.203	0.234	0.245	0.296	0.197	0.2	0.221
B1	0.176	0.153	0.165	0.181	0.227	0.251	0.178	0.169	0.173	0.204	0.223	0.256	0.204	0.188	0.202
C1	0.202	0.172	0.154	0.158	0.194	0.214	0.203	0.172	0.17	0.175	0.216	0.227	0.237	0.202	0.194
D1	0.232	0.204	0.173	0.151	0.161	0.187	0.241	0.21	0.175	0.165	0.17	0.203	0.241	0.233	0.2
E1	0.265	0.227	0.2	0.164	0.163	0.16	0.269	0.237	0.21	0.173	0.162	0.166	0.267	0.241	0.222
F1	0.285	0.268	0.238	0.199	0.166	0.156	0.286	0.27	0.237	0.204	0.17	0.163	0.278	0.269	0.242
A2	0.167	0.172	0.195	0.219	0.242	0.273	0.154	0.157	0.189	0.227	0.257	0.263	0.164	0.171	0.196
B2	0.17	0.164	0.16	0.181	0.231	0.223	0.17	0.156	0.159	0.202	0.224	0.259	0.174	0.164	0.17
C2	0.202	0.179	0.16	0.164	0.182	0.216	0.206	0.173	0.149	0.156	0.191	0.216	0.201	0.18	0.165
D2	0.23	0.208	0.17	0.159	0.168	0.184	0.232	0.203	0.167	0.156	0.156	0.188	0.234	0.204	0.168
E2	0.263	0.234	0.202	0.167	0.164	0.167	0.272	0.227	0.202	0.167	0.158	0.167	0.268	0.232	0.204
F2	0.286	0.26	0.236	0.2	0.174	0.164	0.268	0.259	0.221	0.2	0.165	0.156	0.289	0.27	0.228
A3	0.189	0.193	0.2	0.219	0.246	0.274	0.17	0.172	0.188	0.228	0.242	0.267	0.153	0.168	0.186
B3	0.199	0.194	0.186	0.194	0.232	0.26	0.17	0.167	0.163	0.199	0.214	0.24	0.164	0.155	0.166
C3	0.213	0.199	0.197	0.198	0.197	0.228	0.2	0.171	0.164	0.162	0.186	0.214	0.197	0.168	0.156
D3	0.227	0.205	0.196	0.19	0.184	0.196	0.234	0.204	0.174	0.157	0.164	0.188	0.23	0.191	0.163
E3	0.271	0.236	0.204	0.191	0.187	0.194	0.26	0.238	0.203	0.173	0.163	0.173	0.26	0.225	0.196
F3	0.293	0.259	0.257	0.225	0.206	0.198	0.284	0.257	0.242	0.204	0.177	0.16	0.275	0.249	0.231
A4	0.227	0.224	0.228	0.226	0.261	0.272	0.192	0.186	0.199	0.227	0.245	0.307	0.161	0.172	0.187
B4	0.227	0.215	0.218	0.234	0.25	0.259	0.206	0.186	0.187	0.197	0.226	0.244	0.176	0.169	0.166
C4	0.242	0.216	0.216	0.216	0.22	0.242	0.215	0.2	0.184	0.187	0.204	0.237	0.2	0.176	0.157
D4	0.26	0.247	0.214	0.205	0.21	0.223	0.25	0.219	0.199	0.187	0.19	0.205	0.23	0.206	0.17
E4	0.285	0.247	0.24	0.212	0.222	0.21	0.264	0.223	0.225	0.204	0.188	0.198	0.259	0.23	0.199
F4	0.308	0.285	0.25	0.236	0.227	0.223	0.292	0.271	0.244	0.222	0.195	0.188	0.276	0.258	0.241
A5	0.24	0.244	0.257	0.251	0.264	0.284	0.219	0.23	0.228	0.241	0.273	0.279	0.19	0.186	0.204
B5	0.241	0.234	0.227	0.258	0.25	0.278	0.225	0.221	0.222	0.219	0.231	0.268	0.191	0.186	0.196
C5	0.242	0.25	0.242	0.251	0.237	0.258	0.242	0.224	0.221	0.219	0.221	0.246	0.219	0.194	0.182
D5	0.279	0.262	0.246	0.229	0.244	0.242	0.264	0.235	0.221	0.213	0.222	0.224	0.241	0.219	0.2
E5	0.286	0.277	0.259	0.257	0.233	0.225	0.287	0.265	0.224	0.231	0.208	0.213	0.267	0.236	0.23
F5	0.301	0.289	0.264	0.276	0.249	0.24	0.298	0.292	0.25	0.252	0.228	0.225	0.292	0.263	0.251

282

1st/2nd	D3	E3	F3	A4	B4	C4	D4	E4	F4	A5	B5	C5	D5	E5	F5
A1	0.229	0.269	0.282	0.215	0.235	0.258	0.273	0.283	0.305	0.248	0.27	0.283	0.309	0.326	0.33
B1	0.212	0.243	0.24	0.23	0.222	0.228	0.243	0.259	0.3	0.25	0.253	0.255	0.288	0.28	0.278
C1	0.206	0.216	0.241	0.253	0.226	0.229	0.225	0.237	0.249	0.27	0.259	0.271	0.256	0.255	0.279
D1	0.193	0.193	0.219	0.263	0.248	0.231	0.226	0.221	0.228	0.273	0.262	0.256	0.26	0.244	0.271
E1	0.205	0.19	0.202	0.289	0.254	0.245	0.225	0.224	0.227	0.283	0.275	0.279	0.238	0.253	0.249
F1	0.222	0.207	0.196	0.301	0.261	0.249	0.243	0.236	0.236	0.29	0.303	0.303	0.267	0.26	0.233
A2	0.244	0.255	0.276	0.196	0.2	0.221	0.252	0.281	0.261	0.232	0.232	0.241	0.258	0.283	0.31
B2	0.198	0.228	0.264	0.198	0.191	0.191	0.218	0.242	0.26	0.238	0.21	0.228	0.235	0.264	0.269
C2	0.163	0.194	0.222	0.223	0.193	0.204	0.194	0.2	0.236	0.236	0.235	0.221	0.214	0.232	0.26
D2	0.168	0.176	0.205	0.25	0.224	0.203	0.188	0.195	0.217	0.252	0.249	0.229	0.221	0.224	0.229
E2	0.17	0.171	0.174	0.275	0.244	0.226	0.197	0.196	0.205	0.286	0.264	0.242	0.225	0.215	0.236
F2	0.216	0.184	0.168	0.288	0.272	0.236	0.226	0.207	0.192	0.275	0.269	0.266	0.247	0.227	0.216
A3	0.226	0.253	0.284	0.17	0.172	0.201	0.23	0.274	0.273	0.195	0.2	0.232	0.23	0.257	0.275
B3	0.19	0.233	0.243	0.167	0.164	0.173	0.197	0.23	0.264	0.203	0.201	0.189	0.221	0.231	0.253
C3	0.164	0.187	0.212	0.211	0.173	0.164	0.168	0.205	0.226	0.226	0.198	0.195	0.198	0.213	0.235
D3	0.16	0.166	0.19	0.243	0.217	0.174	0.16	0.172	0.186	0.241	0.215	0.194	0.202	0.193	0.222
E3	0.172	0.158	0.166	0.256	0.232	0.21	0.182	0.166	0.176	0.256	0.235	0.217	0.209	0.199	0.197
F3	0.2	0.167	0.147	0.279	0.27	0.229	0.204	0.179	0.168	0.299	0.278	0.238	0.225	0.202	0.204
A4	0.213	0.245	0.272	0.152	0.164	0.195	0.222	0.254	0.277	0.166	0.178	0.21	0.222	0.266	0.277
B4	0.192	0.23	0.233	0.174	0.155	0.164	0.193	0.216	0.237	0.177	0.166	0.173	0.186	0.222	0.261
C4	0.162	0.191	0.218	0.21	0.173	0.157	0.159	0.187	0.218	0.194	0.183	0.173	0.165	0.197	0.201
D4	0.159	0.16	0.194	0.226	0.199	0.171	0.151	0.156	0.187	0.231	0.216	0.177	0.162	0.172	0.197
E4	0.174	0.16	0.17	0.251	0.224	0.204	0.168	0.154	0.163	0.257	0.243	0.205	0.178	0.164	0.18
F4	0.203	0.171	0.162	0.285	0.25	0.239	0.197	0.174	0.155	0.276	0.267	0.238	0.22	0.184	0.168
A5	0.226	0.266	0.283	0.161	0.169	0.191	0.225	0.25	0.277	0.146	0.159	0.184	0.226	0.262	0.279
B5	0.208	0.227	0.238	0.176	0.161	0.167	0.19	0.214	0.235	0.164	0.156	0.161	0.197	0.226	0.242
C5	0.191	0.202	0.224	0.201	0.17	0.156	0.172	0.194	0.22	0.204	0.172	0.152	0.166	0.19	0.217
D5	0.192	0.184	0.202	0.227	0.212	0.181	0.16	0.168	0.195	0.234	0.198	0.168	0.158	0.162	0.196
E5	0.201	0.19	0.196	0.27	0.225	0.204	0.178	0.162	0.167	0.26	0.235	0.2	0.167	0.153	0.165
F5	0.224	0.196	0.192	0.278	0.269	0.238	0.217	0.177	0.163	0.288	0.262	0.228	0.214	0.163	0.154

After removing outliers, to obtain an entry for a given bi-tap in the empirical bi-tap table representative of a generic user, one can no longer simply take the mean of the remaining data points. This would not give equal weight to each of the users. For example, if there is one outlier in the set of 25, belonging to, say, Participant 3, then if we define the empirical bi-tap table entry for that bi-tap as the mean of the remaining 24 points, Participant 3, having contributed only 4 points to the mean, will be underrepresented by a factor 4/5.

The empirical bi-tap table entries are instead calculated giving equal weight to each user: For a given bi-tap, take the mean time for each of the users on the points remaining in their quintuple after removing outliers, then define the empirical bi-tap table entry to be the average of those five means.

Formally, we calculate the entries as follows. First, we calculate individual bi-tap tables E_s for each test participant s, and then define as the mean of these tables:

$$\bar{E}(k_1, k_2) = \frac{\sum_{s \in S} E_s(k_1, k_2)}{|S|} \qquad (2)$$

where k_1, k_2 are keys, S is the set of test participants, and $|S|$ is the cardinality of S (in our case, 5). The (k_1, k_2)th entry $E_s(k_1, k_2)$ of the individual bi-tap table E_s of test participant $s \in S$ is defined as the mean of the data points for s that remain for the bi-tap (k_1, k_2) after removing outliers using the procedure outlined earlier.

3.4. Interpreting the Empirical Data

Figure 5 is a scatterplot of bi-tap time against bi-tap length for the 900 bi-tap times in the empirical bi-tap table $\bar{E}$ (Figure 4). Distance is measured in key widths. As one would expect, bi-tap time increases with distance.

The vertical spread of the clusters (a cluster is the set of bi-tap times of a given length) is not random. Two patterns are lost in the projection of the data onto a time-distance scatter: Although the time interval between successive taps on a keyboard depends principally on the distance between them, it also depends on *position* and *trajectory*.

These dependencies are illustrated in Figure 6. The left grid shows the three top points of each of clusters *b*, *c*, and *d*, and the right grid shows the bottom three points of the clusters. The fast bi-taps are around the middle columns C and D, heading north/northeast/east; the slow bi-taps are around the left, bottom, and lower boundaries of the grid, heading west/southwest.

Figure 5. (*Upper*) Time-distance scatter plot of the 900 bi-tap times in the empirical bi-tap table (Figure 4), the aggregated results from our experiments. (*Lower*) Chart illustrating a typical relation (modulo rotation and reflection) between keys of bi-taps in clusters *a* to *g* of the plot. For example, cluster *a* consists of double-taps on the same key (30 data points), cluster *b* consists of bi-taps between immediate neighbors (98 data points: 24↑, 24↓, 25→, 25←), cluster *c* consists of bi-taps between diagonal neighbors (80 data points: 20↗, 20↘, 20↙, 20↖), and so on. Cluster means are quoted above the corresponding pictures (.154, .165, ...).

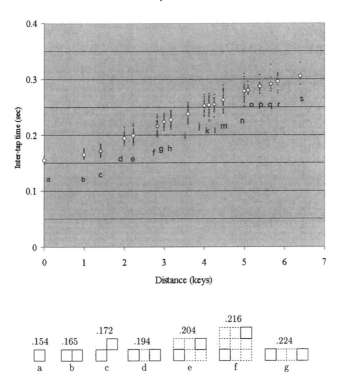

3.5. Relation With Fitts' Law

Fitts' law is a well-known model of human movement (Fitts, 1954; MacKenzie, 1991) that has been used in a number of papers on stylus keyboards (Hunter et al., 2000; Lewis et al., 1999; MacKenzie et al., 1994; MacKenzie & Zhang, 1999; MacKenzie et al., 1999; Soukoreff & MacKenzie, 1995) in the following form:

Figure 6. Fast and slow bi-taps, showing how bi-tap time depends not only on the distance between source and target but also on position and trajectory. The left grid shows the top three points of each of clusters *b, c,* and *d* of Figure 5 (time-distance scatter plot), and the right grid shows the bottom three points of each of *b, c,* and *d.*

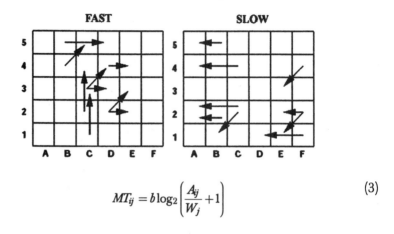

$$MT_{ij} = b \log_2 \left(\frac{A_{ij}}{W_j} + 1 \right) \tag{3}$$

where

MT_{ij} = mean time to move from key i to key j (in seconds)
W_j = size of key j
A_{ij} = distance from key i to key j
b = 1/4.9 is a fitted constant (MacKenzie, Sellen, & Buxton, 1991).

Fitts' law is somewhat inaccurate at a small scale (see, e.g., Section 2.2 of MacKenzie, 1991). This observation is confirmed by the data collected in our experiment: Figure 7 is the time-distance scatter of our empirical bi-tap table, with Fitts' law (in the aforementioned form) superimposed. Another difference between Fitts' law and our empirical results is the dependency of stylus dexterity on position and trajectory, as depicted in Figure 6. Fitts' law, as applied to a rectangular grid of square keys, is a translation and direction invariant.

3.6. Peak Input Rates of Stylus Keyboards

In this section, we use the approach described in Section 2 to predict peak expert input rates on various stylus keyboards. We begin with a simple illustrative example, an ABC Keyboard.

Define K_{ABC} to be the ABC keyboard layout depicted in Figure 8. (Note that keys D1, E1, and F1 are unused.) With the bi-gram probability table of the Appendix and the empirical bi-tap table (Figure 4), our model predicts the following peak expert text-input rate for the ABC layout:

Figure 7. Relation between Fitts' law and our empirical data.

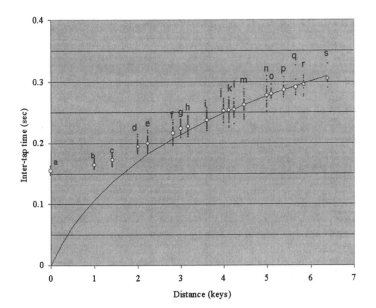

Figure 8. Four keyboard layouts. Shaded keys are unused keys.

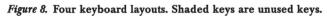

Q	F	U	M	C	K	Z
SPACE		O	T	H	SPACE	
B	S	R	E	A	W	X
SPACE		I	N	D	SPACE	
J	P	V	G	L	Y	

OPTI

Z	V	C	H	W	K
F	I	T	A	L	Y
SPACE		N	E	SPACE	
G	D	O	R	S	B
Q	J	U	M	P	X

FITALY

A	B	C	D	E	F
G	H	I	J	K	L
M	N	SPACE		O	P
Q	R	S	T	U	V
	W	X	Y	Z	

ABC-center

A	B	C	D	E	F
G	H	I	J	K	L
M	N	O	P	Q	R
S	T	U	V	W	X
Y	Z	SPACE			

ABC

287

$$R(K_{ABC}, \bar{B}, \bar{E}) = 4.702\text{char}/\sec = 56.42\text{wpm}$$

where wpm denotes words per minute, with a word defined as 5 characters (including Spaces).

3.7. Validation of the Bi-Tap Table

In this section we compare our peak input rate predictions for two of the keyboard layouts shown in Figure 8 (the OPTI and the FITALY) with previous measures of peak input rate, thus validating the user model described by our empirical bi-tap table.

The OPTI Validation

In MacKenzie and Zhang (1999), users were trained over 20 sessions to tap 70 stock phrases on the OPTI layout depicted in Figure 8. Average text input rates followed the power law of learning from 17 wpm (1.42 char/sec) in Session 1 to 44.2 wpm (3.7 char/sec) in Session 20, and a regression ($R^2 = .997$) predicted a performance of 60.7 wpm (5.06 char/sec) on Session 50.

As can be seen in Figure 9, our predictions of peak text input rate correspond nicely with the asymptotic predictions of MacKenzie and Zhang. $\text{OPTI}_{\text{lower right}}$ denotes use of the OPTI with a fixed choice of Space key as the lower right of the four alternatives, and $\text{OPTI}_{\text{last}}$ denotes use of the OPTI where the user chooses the Space key closest to the last letter tapped. The details of the calculations, including the construction of the 5×7 bi-tap table, can be found in Section 3.12.

The FITALY Validation

The FITALY (Textware Solutions, 2000) is a commercially available stylus keyboard, the layout of which is shown in Figure 8. Figure 10 reproduces the results of a promotional competition held by the manufacturers in which contestants were timed tapping the following 181-character paragraph on the FITALY:

What you need to do to have a chance to win the contest is to tap this sentence as fast as you can without any error. One more thing you need to have for a valid entry is a witness.

Our predictions of peak expert text input rate are interleaved in the table (Figure 10). Video recordings of human performance (available on the

Figure 9. Validating our bi-tap table by comparing peak expert text input rate predictions with results of MacKenzie and Zhang (1999) on the OPTI keyboard depicted in Figure 8. See Section 3.11 for details of the calculations.

	Characters/Second	Words Per Minute
MacKenzie and Zhang Session 50	5.06	60.7
$R(\text{OPTI}_{\text{lower right}},\ \bar{B},\ \bar{E}_7)$	5.10	61.2
$R(\text{OPTI}_{\text{last}},\ \bar{B}^*,\ \bar{E}_7)$	5.25	63.0

Figure 10. Validating our peak expert text input rate predictions using the commercial FITALY keyboard, the layout of which is depicted in Figure 8. The table is reproduced from the FITALY Web site (Textware Solutions, 2000), together with the interleaving of our predictions R_q (for q indicating various different patterns of Space key choice, detailed in Section 3.13) of peak expert text input rates. Shown are the performances of the top 10 competitors in a June to July 2000 speed-tapping competition. The competition task was to tap the 181-character paragraph quoted in Section 3.7.

Competitor/Prediction	Characters/Second	Words Per Minute
1	6.165	73.98
2	5.803	69.64
3	5.798	69.47
4	5.607	67.38
R_{best}	5.349	64.18
R_{last}	5.329	63.95
5	5.288	63.46
R_{right}	5.255	63.06
R_{random}	5.223	62.68
6	5.191	62.29
R_{left}	5.189	62.27
7	5.130	61.56
8	4.805	57.66
9	4.756	57.07
10	4.694	56.33

Note. Char/sec = characters per second; wpm = words per minute.

FITALY Web site) are of individuals remarkably highly trained on entering this 181-character sequence, and as such represent peak input speeds. The fact that our predictions lie within the table adds to the validation of the bi-tap table for measuring peak input speed. Note that the sample test for the competition contains punctuation: two periods and two capitalizations. Hence, our predictions will be marginally too high. However, they still fall in essentially the same positions within Figure 10.

3.8. Stability With Respect to Bi-Gram Table

Our chosen bi-gram probability table is a composite of three bi-gram frequency tables B_1, B_2, and B_3. These three tables are reproduced in the Appendix. B_1 is Soukoreff and MacKenzie's (1995) extension of Mayzner and Tresselt's (1965) 26 × 26 table to include the Space character. B_2 is the 26 × 26 bi-gram table of Konheim's (1981) introductory cryptography textbook, to which we have added the same space bi-gram extension.[2] There are discrepancies between B_1 and B_2, possibly due to the fact that they were built from small text corpora. The following discrepancies[3] are the most notable: OF (80 vs. 731), ON (598 vs. 1232), TI (252 vs. 865), OU (1115 vs. 533), TH (3774 vs. 2161), HE (3155 vs. 2053).

To reduce these discrepancies we created a third bi-gram table B_3 of our own from a corpus 10 times the size, a mixture of informal and formal English (e-mail and classic novels). Stop-lists were used on proper nouns, and so forth. See Manning and Schütze (1999) for techniques for sampling data from text corpora. Then we defined $\bar{B}$ as the normalization of the weighted mean of (appropriate rescalings of) B_1, B_2, and B_3 (see Appendix). Figure 11 shows predictions of peak expert text input rate for the ABC keyboard with the four choices of bi-gram table. The observed rates did not vary significantly with selection of the bi-gram table.

3.9. Optimizing Stylus Keyboard Layout

The problem of finding the optimal layout for a stylus keyboard is equivalent to the problem of minimizing the average time between tapping two keys. Recall from Section 2 that the average time between tapping two keys on keyboard layout K with input text modeled by the bi-gram probability table P and stylus dexterity modeled by the bi-tap table E, is

$$\bar{M}(K,P,E) = \sum_{\alpha,\beta} P(\alpha,\beta)E(K(\alpha),K(\beta)) \qquad (4)$$

where α and β range over the character set. Fixing $P = \bar{B}$ (see Appendix) and $E = \bar{E}$ (the empirical bi-tap table; Figure 4), our task is to minimize this expression with respect to K, a function from characters to keys. This has the

---

2. Because Konheim's total count of A to Z bi-grams (67,227) is nearly identical to that of Mayzner and Tresselt (67,320), we can conveniently add the Space bi-grams with only minor renormalization.

3. Note that direct comparisons of bi-gram frequencies between the tables makes sense because the sums of entries are nearly identical.

Figure 11. Stability of peak expert text input rate prediction with respect to the bi-gram table B, with the ABC keyboard. $\bar{E}$ is the empirical bi-tap table shown in Figure 4.

Bi-gram table B	$R(\text{ABC}, B, \bar{E})$	
	char/sec	wpm
$\bar{B}$	4.702	56.42
B_1	4.691	56.28
B_2	4.693	56.32
B_3	4.723	56.68

A	B	C	D	E	F
G	H	I	J	K	L
M	N	O	P	Q	R
S	T	U	V	W	X
Y	Z	SPACE			

form of a standard optimization problem called the *quadratic assignment problem* (QAP).

The QAP (Koopmans & Beckman, 1957) has been shown to be an extremely hard problem. Not only is it NP-hard (Sahni & Gonzalez, 1976), but it is NP-hard to approximate its optimal solution to within any constant factor (Queyranne, 1986). There are a number of heuristics that can be employed to find reasonable solutions to the QAP: genetic algorithms, the Metropolis method, and dynamic simulation, to name but a few. We chose to implement the hybrid ant system proposed by Gambardella, Taillard, and Dorigo (1997) because it has been shown to find quality solutions quickly.

The best solution found by the hybrid ant system was the keyboard layout $K_{\bar{B},\bar{E}}$ depicted in Figure 12, with a predicted peak expert text input rate of

$$R(K_{\bar{B},\bar{E}}, \bar{B}, \bar{E}) = 5.438 \,\text{char/sec} = 65.26 \text{ wpm}.$$

See Section 3.11 for a comparison with other keyboards.

3.10. Variation of Best Solution With Respect to Bi-Gram Table

Figure 13 shows how the best solution produced by the hybrid ant system varies with respect to the bi-gram table parameterizing the optimization problem. One can observe how the final layout is directly related to the idiosyncrasies of a particular bi-gram table. For example, recall the major discrepancies between the frequency tables B_1 (Mayzner and Tresselt) and B_2 (Konheim): OF (80 vs. 731), ON (598 vs. 1232), TI (252 vs. 865), OU (1115 vs. 533), TH (3774 vs. 2161), HE (3155 vs. 2053). Notice how the strong preference of B_2 for

Figure 12. The keyboard layout $K_{\bar{B},\bar{E}}$, the best solution found by the hybrid ant system to the quadratic assignment problem of optimal stylus keyboard layout. The predicted peak expert text input rate $R(K_{\bar{B},\bar{E}}, \bar{B}, \bar{E})$ is 5.438 char/sec = 65.26 wpm.

K	G	I	C	Z
F	N	T	H	W
O	S	SPACE	A	Y
U	R	E	D	V
P	M	L	B	X

OF, ON, and TI over B_1 (by an order of magnitude in the case of OF) is observable in the layouts: the keys of the pairs OF and IT are directly adjacent, and the keys O and N of ON are diagonally adjacent. These three bi-grams stretch over distances of 3.2, 2, and 3.2 key widths, respectively. Figure 14 shows the variation in peak input speed of the four optimized layouts under evaluation with respect to each of the four bi-gram tables.

3.11. Benchmarking Various Keyboards

Using our predictions of peak expert text input rate, we benchmark four keyboard layouts against the simple ABC layout: the OPTI, the FITALY, the best solution $K_{\bar{B},\bar{E}}$ to the keyboard layout optimization problem discovered by the hybrid ant system, and a variant of the ABC with Space at the center, which we call ABC-center. The five layouts in question are depicted in Figures 8 and 12.

The results are shown in Figure 15. Observe that simply moving the Space key to the center of the ABC already increases performance by nearly 4%. The FITALY yields an additional 9.4% increase relative to the ABC. Our ant algorithm solution $K_{\bar{B},\bar{E}}$ gains yet another 2.3% above the FITALY. These results are not definitive because a sample size of only 5 users was used to generate our table of bi-tap data.

In the following sections we report in detail the calculations involved in doing the simulations for the FITALY and OPTI. They were nontrivial because the layouts have multiple Space keys.

3.12. Predicting the Peak Expert Text Input Rate of the OPTI

Concerning their OPTI layout, MacKenzie and Zhang (1999) noted that "having four SPACE keys is convenient; but, using the optimal SPACE key re-

Figure 13. Variation of best solution found by the hybrid ant system optimization with respect to the bi-gram probability table.

	K	G	I	C	Z
	F	N	T	H	W
Q	O	S	SPACE	A	Y
J	U	R	E	D	V
	P	M	L	B	X

Layout $K_{\bar{B},\bar{E}}$

$R(K_{\bar{B},\bar{E}},\bar{B},\bar{E}) = 5.438$ char/sec

	Z	F	C	U	Q
	V	O	I	N	G
J	M	R	T	A	L
	P	E	SPACE	S	Y
X	K	H	D	W	B

Layout $K_{B_1,\bar{E}}$

$R(K_{B_1,\bar{E}},B_1,\bar{E}) = 5.476$ char/sec

	Q	P	Y	U	J
	C	S	T	O	B
K	L	H	SPACE	R	M
X	I	A	E	D	F
Z	V	W	N	G	

Layout $K_{B_2,\bar{E}}$

$R(K_{B_2,\bar{E}},B_2,\bar{E}) = 5.454$ char/sec

	W	G	I	C	Z
	D	N	T	H	K
F	O	SPACE	E	S	V
J	U	R	A	L	X
Q	P	M	Y	B	

Layout $K_{B_3,\bar{E}}$

$R(K_{B_3,\bar{E}},B_3,\bar{E}) = 5.432$ char/sec

Figure 14. Stability of the peak expert text input rate with respect to choice of bi-gram table. Entries in the table are in words per minute. The row label is the bi-gram table used in the ant algorithm search; the column label indicates the bi-gram table used in the peak expert text input rate prediction *R*. Thus each row consists of four different evaluations of a keyboard.

Bi-gram Table Used for Ant Algorithm	Bi-gram Table Used for Prediction Rates			
	B1	B2	B3	$\bar{B}$
B1	65.486 wpm	64.709 wpm	64.707 wpm	64.935 wpm
B2	64.650 wpm	65.443 wpm	64.643 wpm	64.869 wpm
B3	65.112 wpm	65.123 wpm	65.279 wpm	65.137 wpm
$\bar{B}$	65.265 wpm	65.345 wpm	65.202 wpm	65.242 wpm

Note. wpm = words per minute.

quires extra judgment on-the-fly and this is not likely to occur—at least within the confines of the limited practice in this study."

Hence, for our predictions, we considered the two patterns of Space use observed by MacKenzie and Zhang in the penultimate section of their paper: (a)

Figure 15. Benchmarking results for the five keyboard layouts depicted in Figures 8 and 12. The right column shows the percentage performance increase above the ABC. The subscripts on the FITALY and OPTI indicate the pattern of space usage (see Section 3.7 for details). Because on-the-fly calculations of optimal tri-character path are somewhat unrealistic to expect from users (see MacKenzie & Zhang, 1999), we suggest FITALY$_{last}$ as the most representative of FITALY performance. Predictions of peak expert input speeds are not estimates of real continuous performance (see Section 2.1).

	Predicted Peak Expert Text Input Rate		
Keyboard K	Char/sec	wpm	% > ABC
K $_{\bar{B},\bar{E}}$	5.438	65.26	15.65
FITALY$_{best}$	5.349	64.19	13.77
FITALY$_{last}$	5.329	63.95	13.35
FITALY$_{right}$	5.255	63.06	11.77
OPTI$_{last}$	5.248	62.98	11.63
FITALY$_{random}$	5.223	62.67	11.08
FITALY$_{left}$	5.189	62.27	10.37
OPTI$_{lower\ right}$	5.103	61.24	8.54
ABC-center	4.888	58.65	3.95
ABC	4.702	56.42	0.00

Note. char/sec = characters per second; wpm = words per minute.

the closest Space key to the last character tapped, denoted OPTI$_{last}$; and (b) the lower right Space key (of the four, chosen because right-handers may be naturally inclined to pick the Space key that obscures their view of the grid the least), denoted OPTI$_{lower\ right}$.

Our empirical bi-tap table $\bar{E}$ (Figure 4) was obtained for a 5 × 6 grid, but the OPTI has a 5 × 7 grid. Rather than repeating our experiments in full with a 5 × 7 grid, we constructed a 5 × 7 empirical bi-tap table $\bar{E_7}$ from $\bar{E}$ as follows:

- The bi-tap values in the left 5 × 6 subgrid of $\bar{E_7}$ are taken directly from $\bar{E}$.
- Making the simplifying assumption that the inner half of each Space key is used, this leaves only two keys unaccounted for in column 7—Z and X.
- The bi-gram probabilities between characters of column 1 (Q, B, J) and column 7 (Z, X) are all zero; hence, one can assign arbitrary times to the bi-taps of $\bar{E_7}$ between column 1 and column 7 (we chose 0 sec) without any effect on the model.
- This leaves the bi-tap times between Z and X and columns 2–7. Because this is a 5 × 6 subgrid, we used $\bar{E}$ (shifted right by one column).

In essence, we are thinking of the OPTI as the 5×6 grid consisting of columns 1–6, with the infrequently used X and Z "pasted" on the side. Note that the use of this artificially constructed 5×7 table in peak expert input speed prediction should be reasonably accurate because the only bi-grams that are not covered by the 5×6 case are the 0 probability pairings between {Q, B, J} and {Z, X}.

OPTI Prediction: Lower Right Space Choice

The prediction of peak expert text input rate $R(\text{OPTI}_{\text{lower-right}}, \bar{B}, \bar{E}_7)$ in Figure 9 was based on the following data. The character set was $C_{27} = \{\text{A–Z}, \text{Space}\}$. The bi-gram probability table $\bar{B}$ is from the Appendix, and the bi-tap table $\bar{E}_7$ is that derived from $\bar{E}$ (Figure 4) in the manner described earlier.

OPTI Prediction: Closest-to-Last Space Choice

The prediction of peak expert text input rate $R(\text{OPTI}_{\text{last}}, \bar{B}, \bar{E}_7)$ in Figure 9 was based on the following data. The keyboard function $\text{OPTI}_{\text{last}}$, assigning keys to characters, was the same as $\text{OPTI}_{\text{lower right}}$ on characters A to Z, and with the following assignment of Space characters (lower left, lower right, upper left, upper right) to keys:

SPACE_{ll}	SPACE_{lr}	SPACE_{ul}	SPACE_{ur}
B2	F2	B4	F4

where we use the key-naming convention of Figure 3 with an additional right-hand column G1 to G5. The bi-tap table $\bar{E}_7$ is that derived from $\bar{E}$ in the manner described at the beginning of Section 3.12.

The bi-gram probability table $\bar{B}$ used is $\bar{B}$ on A to Z, together with bi-grams involving the four Space characters calculated as follows:

- The probability of consecutive spaces is zero.
- Given an A to Z character α and one of the four spaces σ, the probability (α, σ) of the bi-gram (α, σ) is (α, Space) if the σ key is the closest space to the α key on the grid (picking right-most and lowest in case of tie-break), and 0 otherwise.
- Given an A to Z character α and one of the four spaces σ, the probability $\bar{B}$ of the bi-gram (α, σ) is $\theta_\sigma \bar{B}(\text{Space}, \alpha)$, where θ_σ is the proportion of the time σ occurs as a trailing space:

$$\theta_\sigma = \frac{\sum_{\beta \in \{A,...,Z\}} \bar{B}_*(\beta, \sigma)}{\sum_{\beta \in \{A,...,Z\}} \sum_{X=ll,lr,ul,ur} \bar{B}_*(\beta, \text{Space}_X)} \tag{5}$$

3.13. Predicting the Peak Expert Text Input Rate of the FITALY

The cases FITALY_{last}, FITALY_{right}, and FITALY_{left} in Figure 15 are analogous to OPTI_{last}, $\text{OPTI}_{lower\ right}$, and $\text{OPTI}_{lower\ left}$, respectively. FITALY_{random} denotes a random choice of Space key: We used the bi-gram table derived from that splits the probability of bi-grams of $\bar{B}$ involving Space equally in two.

The case FITALY_{best} was labor intensive. We considered by hand all 676-character tri-grams of the form $(\alpha, \text{Space}, \beta)$ for alphabet characters α and β, and decided which of the two Space keys would be the optimal choice. Having assigned left/right to each such tri-gram, the probability of the bi-gram $(\alpha, \text{Space}_{left})$ is

$$\bar{B}(\alpha, \text{Space}) = \frac{\sum_{B \in L(\alpha, \beta)} \bar{B}(\text{Space}, \beta)}{\sum_\beta \bar{B}(\text{Space}, \beta)} \tag{6}$$

where β ranges over the character set, and $L(\alpha, \beta)$ is the subset of characters for which the tri-gram $(\alpha, \text{Space}, \beta)$ was designated as using the left Space key. The other space bi-gram probabilities work similarly.

Note that, in the absence of tri-gram data, one has to assume a uniform distribution of tri-grams with respect to bi-grams. The effect of this assumption should be negligible.

3.14. Stylus Keyboards: Conclusion

We illustrated the empirical bi-action table technique in the context of stylus keyboard design. A hybrid ant system yielded an optimized stylus keyboard layout (Figure 12) that outperformed the commercial FITALY layout. This result, however, is not to be considered final due to the small number of 5 participants used to generate our table of bi-tap data and the coincidence of the training data and test data.

There is an observation that stems from our experiments: All previous related work on stylus keyboard design has only considered distance between two keys as a predictor of the duration of the motion between these keys. We found that this duration depends also on the first key position and on the relative position of the second key (see Figure 6). These dependencies, and any other more subtle dependencies that could be impossible to model, are automatically taken into account by the empirical bi-action table.

Within the realm of stylus keyboard design, the following topics are possibilities for future work:

- *Bi-tap corpus:* Although 25 data points for each of 900 bi-taps is a considerable amount of data, a clear route to improving our work would be to obtain a much larger corpus of test data. Just as bi-gram tables based on large corpora of text are publicly available to those who which to use it (e.g., in cryptology or natural language processing), it would make sense to make publicly available a bi-tap table built on a considerable corpus of stylus dexterity test data. Researchers interested in benchmarking and optimizing their own layouts (e.g., with a variety of different character sets, perhaps from a variety of languages) could make use of the corpus bi-tap table.

- *Errors:* It would be nice to incorporate a quantitative account of errors.

- *Obscuration:* As a right-hander, having just tapped a key in the top left of the grid, when aiming for a key in the mid- or lower right, is there any delay due to the fact that my hand is obscuring the target zone?

- *Investigate position dependency:* A possible explanation for the dependency of bi-tap time on position (preference for the center; see Figure 6) is that the users rest on the side of the PDA with the wrist, outer-palm, and/or outer edge of little finger. The natural resting position of the stylus is somewhere over the center of the grid. Motions around the edges of the grid require either a cramped or overstretched position of the fingers and thumb, or a cocking of the wrist.

- *Investigate trajectory dependency:* A possible explanation for the dependency of bi-tap times on trajectory (dislike for heading west; see Figure 6) is that everyday handwriting is from left to right. Low-level motor skills in the hand involved in moving against the direction of writing are probably less well developed. To further investigate trajectory dependency one could carry out the tests with left-handers and/or people whose mother tongue is written from right to left.

- *Optimization incorporating double-Space and/or double-E:* The problem of finding the optimal keyboard with two Space keys and/or two E keys is much harder than the quadratic assignment problem. It would be useful to determine whether having two Space keys can yield a faster layout.

- *Tri-taps:* Although the extent of dependency of bi-taps on the preceding tap is probably low, it may be useful to collect data for an empirical tri-tap table. However, with 27,000 possible triples (in the case of 30 keys), experiments would be impractical.

- *Keyboard shape:* Hunter et al. (2000) and Zhai et al. (2000) considered layouts on a hexagonal grid. One could also imagine a radial pattern, a dartboard shape with Space as the "bull's-eye." Empirical bi-tap tables could be obtained for these shapes and used to produce optimized key layouts. One observation of a rectangular grid arrangement is that it maximizes the number of adjacent keys. Define a *neighbor* of a key to be any other key that is reachable without having to leap over an intermediate key. In a hexagonal grid, each key has only six neighbors, whereas in a rectangular grid each key has eight. Our empirical data (Figure 5) show an observable jump between neighbor bi-tap times (clusters *a*, *b*, and *c*) and times of bi-taps involving leaps (cluster *d* and beyond). This neighborhood property may be particularly important with regard to the number of neighbors of the central Space key, by far the most frequently used key, in which case a rectangular arrangement might confer an advantage over a hexagonal pattern.
- *x/y-Scale:* The commercial version of the FITALY has rectangular keys that are longer in the horizontal direction. Does such a feature speed up or slow down text input?

4. CONCLUSION

In this article we presented a technique for predicting peak expert input speeds on text input systems. The technique is intended as a tool in the interface development cycle between initial evaluations using abstract mathematical models (e.g., Fitts' law, Hick's law, and the power law) and final evaluations by full empirical testing. We illustrated the approach in the context of stylus keyboards. Empirical bi-action tables could be used in the analysis and design of a wide variety of text input systems, such as two-handed keyboards, chording keyboards, cell phones, glove gesture input, and forms of stylus input including Graffiti, Quikwriting (Perlin, 1998), and Unistrokes (Goldberg & Richardson, 1993).

Relative to full empirical testing, the independence of the empirical bi-action table *E* from the logical aspect of the input system (the character map) confers the following advantages: (a) A change in character mapping (e.g., a change in a keyboard layout) does not demand fresh experimental trials; (b) we avoid the cost of training participants up to expert level with a particular character mapping (e.g., keyboard layout); and (c) having obtained *E*, we can perform an algorithmic search for optimal character mappings. One disadvantage is the reduction in accuracy due to the higher level of abstraction.

Relative to mathematical modeling with laws we cite two advantages: (a) greater generality in the sense of coverage of the full range of input systems, including those for which laws do not easily apply (e.g., glove gestures) and (b) greater accuracy due to the higher specificity of empirical testing. Disadvantages include the cost of empirical testing and the necessity of undertaking new tests for each new input system, aside from the case of variations in character map.

We have not discussed issues such as ease of use or repetitive strain injury. The appropriate balance between such issues and the optimization of input speed should be borne in mind by any researcher choosing to employ our technique.

NOTES

Acknowledgments. We acknowledge Julien Basch, Vaughan Pratt, the participants of our experiments, and funding from the Stanford Wearable Computing Laboratory. We are appreciative of extremely insightful feedback from Scott MacKenzie and three anonymous referees.

Authors' Present Addresses. Dominic Hughes, Gates Building, Room 486, Computer Science Department, Stanford University, Stanford, CA 94305. E-mail: dominic@cs.stanford.edu. James Warren, Gates Building, Room 118, Computer Science Department, Stanford University, Stanford, CA 94305. E-mail: warren@sccm.stanford.edu. Orkut Buyukkokten, Gates Building, Room 430, Computer Science Department, Stanford University, Stanford, CA 94305. E-mail: orkut@stanford.edu.

HCI Editorial Record. First manuscript received November 6, 2000. Accepted by I. Scott Mackenzie. Final manuscript received May 29, 2001. — *Editor*

REFERENCES

Card, S., Moran, T., & Newell, A. (1983). *The psychology of human–computer interaction.* Hillsdale, NJ: Lawrence Erlbaum Associates, Inc.

Fitts, P. (1954). The information capacity of the human motor system in controlling the amplitude of movement. *Journal of Experimental Psychology, 47,* 381–391.

Gambardella, L., Taillard, D., & Dorigo, M. (1999). Ant colonies for the QAP. *Journal of the Operational Research Society, 50,* 167–176.

Goldberg, D., & Richardson, C. (1993). Touch-typing with a stylus. *Proceedings of INTERCHI 93 Conference on Human Factors in Computing Systems.* New York: ACM.

Hunter, M., Zhai, S., & Smith, B. (2000). Physics-based graphical keyboard design (interactive poster). *Proceedings of the CHI 2000 Conference on Human Factors in Computing Systems.*

Konheim, A. (1981). *Cryptography: A primer.* New York: Wiley.

Koopmans, T., & Beckman, M. (1957). Assignment problems and the location of economic activities. *Econometrica, 25,* 53–76.

Lewis, J., LaLomia, M., & Kennedy, P. (1999). Evaluation of typing key layouts for stylus input. *Proceedings of the Human Factors and Ergonomics Society 43rd Annual Meeting.*

MacKenzie, I. (1991). *Fitts' law as a performance model in human–computer interaction.* Unpublished doctoral dissertation, University of Toronto, Canada.

MacKenzie, I., Nonnecke, R., McQueen, J., Riddersma, S., & Meltz, M. (1994). A comparison of three methods of character entry on pen-based computers. *Proceedings of the Human Factors and Ergonomics Society 38th Annual Meeting (Human Factors and Ergonomics Society).*

MacKenzie, I., Sellen, A., & Buxton, W. (1991). A comparison of input devices in elemental pointing and dragging tasks. *Proceedings of the CHI 99 Conference on Human Factors in Computing Systems.*

MacKenzie, I., & Zhang, S. (1999). The design and evaluation of a high performance soft keyboard. *Proceedings of the CHI 99 Conference on Human Factors in Computing Systems.*

MacKenzie, I., Zhang, S., & Soukoreff, R. (1999). Text entry using soft keyboards. *Behaviour & Information Technology, 18,* 235–244.

Manning, C., & Schütze, H. (1999). *Statistical natural language processing.* Cambridge, MA: MIT Press.

Mayzner, M., & Tresselt, M. (1965). Tables of single-letter and digram frequency counts for various word-length and letter-position combinations. *Psychonomic Monograph Supplements, 1*(2).

Noel, R., & McDonald, J. (1989). Automating the search for good designs: About the use of simulated annealing and user models. *Proceedings of Interface 89 (Human Factors Society).*

Perlin, K. (1998). Quikwriting: Continuous stylus-based text entry. *Proceedings of the UIST 98 Symposium on User Interface Software and Technology.*

Queyranne, M. (1996). Performance ratio of heuristics for triangle inequality quadratic assignment problems. *Operations Research Letters, 4,* 231–234.

Sahni, S., & Gonzalez, T. (1976). P-complete approximation problems. *Journal of the ACM, 23,* 555–565.

Soukoreff, R., & MacKenzie, I. (1995). Theoretical upper and lower bounds on typing speed using a stylus and soft keyboard. *Behaviour & Information Technology, 14,* 370–379.

Textware Solutions. (2000). The FITALY one-finger keyboard. Retrieved from http://www.fitaly.com/domperignon/domperignon2.htm

Zhai, S., Hunter, M., & Smith, B. (2000). The Metropolis keyboard: An exploration of quantitative techniques for virtual keyboard design. *Proceedings of the UIST 2000 Symposium on User Interface Software and Technology.*

APPENDIX: BI-GRAM FREQUENCY TABLES

The bi-gram probability table used for peak expert text-input speed predictions is the normalization of the bi-gram frequency table depicted in Figure A–1. The table in Figure A–1 is the equally weighted mean of the (appropriate rescalings of) bi-gram frequency tables B_1, B_2, and B_3 shown in Figures A–2, A–3, and A–4. The composite table in Figure A–1 was constructed to soften the idiosyncrasies of B_1 and B_2, as described in Section 3.8.

Figure A-1. The composite bi-gram frequency table whose normalization *B* was used for predicting peak expert text input rates.

1st/2nd	a	b	c	d	e	f	g	h	i	j	k	l	m	n
a	288	13549	25757	31409	481	5729	12790	815	29360	757	9581	60550	17202	127606
b	10479	899	108	60	40457	12	25	4	6509	1024	0	12961	477	12
c	31031	0	3225	48	35807	42	8	35759	11285	0	12253	9831	132	124
d	11842	183	137	3508	44599	290	1970	52	20818	254	42	2398	1023	847
e	51653	1986	22422	77414	36532	10572	5980	1478	10417	214	2617	32996	20979	87818
f	11893	4	0	17	13550	8257	8	4	18093	0	0	4425	128	52
g	9145	31	25	87	25779	665	5853	19216	5766	0	714	3234	1764	5721
h	88735	155	163	110	232576	38	0	93	62045	0	0	525	392	639
i	11684	3535	36606	25487	19893	7693	18733	39	405	29	4834	29231	21402	137417
j	804	0	0	34	2427	0	0	0	435	0	0	0	0	0
k	710	35	0	43	23129	150	55	230	7929	8	124	783	83	5011
l	31972	407	483	22603	53784	4386	1761	17	35748	0	2604	44528	1842	100
m	35449	5107	232	17	52729	299	4	4	19136	4	4	275	6301	615
n	15884	200	19972	92846	49706	2879	68077	439	19725	598	5391	5501	1592	6121
o	4717	5779	7743	11304	2484	46891	3506	751	7393	405	7666	19636	39612	93263
p	17751	0	25	79	27376	146	29	3311	7246	29	0	14601	540	31
q	0	0	0	0	0	0	0	0	0	93	0	0	0	0
r	35016	1147	4275	13991	114733	1256	5414	1365	38598	51	6719	5794	9389	10096
s	18535	578	6451	505	59388	721	112	24381	29692	59	3867	4172	3570	1115
t	29959	149	2622	70	68067	261	97	262385	56590	0	12	7708	1134	546
u	6068	6778	10510	4137	7637	1031	10742	66	6028	0	157	24086	7017	27413
v	5133	0	225	62	60579	0	0	0	14432	0	0	8	39	0
w	38422	141	21	383	26190	17	0	0	31920	0	56	852	12	7133
x	1672	0	1274	0	1208	21	4	4	52	0	0	8	35	0
y	1023	512	220	189	9478	38	35	35	48	0	31	628	773	364
z	822	0	0	0	2748	0	0	0	0	0	0	232	0	0
SPC	184447	89025	77608	48894	39111	88609	36904	120949	52582	7472	11872	58899	80066	42462

1st\2nd	o	p	q	r	s	t	u	v	w	x	y	z	SPC
a	288	11944	12	70150	63881	91864	7851	19696	4753	1414	23499	1060	20307
b	15192	21	0	7893	2528	966	15334	344	35	0	7624	0	2599
c	42145	0	147	8308	1281	19584	7075	0	0	0	1618	46	4105
d	14685	149	31	6377	7865	185	6373	1111	325	34	4183	0	219394
e	3475	10538	2014	132134	71565	31389	1465	16112	8	10305	11296	278	415298
f	35100	8	0	15032	294	7001	5807	0	31	4	569	0	29183
g	7669	0	4660	6981	3095	2192	3220	0	398	186	300	4	59251
h	36088	62	0	6456	772	12546	4717	0	0	0	2287	0	59925
i	32513	4531	541	20973	65647	65916	504	14844	55	1391	35	2950	406
j	3205	0	0	31	0	0	4668	0	1003	0	4	0	0
k	370	31	0	97	4031	56	149	1471	8	0	530	0	24673
l	25575	1986	0	962	9276	6211	7119	31	434	0	25979	4	53418
m	23368	11325	4	159	4931	114	7520	1929	127	0	7298	0	36885
n	31639	236	322	383	23368	54107	4515	0	17	157	8192	87	113950
o	20460	7962	0	23236	3704	4174	6249	3747	970	0	801	0	11740
p	0	0	0	0	0	0	7923	12	2250	0	0	0	17
q	0	0	0	0	0	0	8180	0	0	0	0	0	0
r	46514	2257	8	8762	26177	20340	16837	101	6798	31	15545	0	122685
s	22918	10336	593	334	22186	70788	12275	114	72	0	2862	0	197773
t	60529	126	43	22881	19760	14855	199	31	0	0	10878	145	205492
u	451	9696	0	35360	27068	35532	96	0	21	169	425	188	22756
v	3132	0	4	4	66	0	79	0	17	0	625	0	97
w	19414	12	0	2420	2822	203	261	0	270	0	131	0	25815
x	79	3132	0	4	0	2371	189	0	0	163	46	0	1944
y	22990	1088	0	296	5802	707	82	0	0	0	62	48	104700
z	207	0	0	0	93	0	0	0	0	8	110	448	235
SPC	84385	52128	4297	42579	139178	330461	16716	11589	150110	106	37956	192	0

Figure A–2. Bi-gram frequency table B_1, Soukoreff and MacKenzie's (1995) extension of Mayzner and Tresselt's (1965) 26 × 26 table to include the Space character. From "Theoretical upper and lower bounds on typing speed using a stylus and soft keyboard," by R. Soukoreff and I. MacKenzie, 1995, *Behaviour & Information Technology, 14.* Copyright 1995 by Taylor & Francis, Ltd. Reprinted with permission.

1st/2nd	a	b	c	d	e	f	g	h	i	j	k	l	m	n
a	2	144	308	382		1	138	9	322	7	146	664	177	1576
b	136	14	0	0	415	0	0	0	78	18	0	98	1	0
c	368	0	13	0	285	0	0	412	67	1	178	108	0	1
d	106	1	0	37	375	0	19	0	148	1	0	22	1	2
e	670	8	181	767	470	0	46	15	127	1	35	332	187	799
f	145	0	0	0	154	0	0	0	205	0	0	69	3	0
g	94	1	0	0	289	0	19	0	96	0	0	55	1	31
h	1164	0	0	0	3155	0	0	1	824	0	0	5	0	0
i	23	7	304	260	189	0	233	288	1	0	86	324	255	1110
j	2	0	0	0	31	0	0	0	9	0	0	0	0	0
k	2	0	0	0	337	0	0	0	127	0	0	10	1	82
l	332	4	6	289	591	0	7	0	390	0	38	546	30	1
m	394	50	0	0	530	0	0	0	165	8	0	4	28	4
n	100	2	98	1213	512	0	771	5	135	3	63	80	0	54
o	65	67	61	119	34	0	9	1	88	0	0	218	417	598
p	142	0	1	0	280	0	0	24	97	0	123	169	0	0
q	0	0	0	0	0	0	0	0	0	0	0	0	0	0
r	289	10	22	133	1139	0	59	21	309	0	53	71	65	106
s	196	9	47	0	626	0	1	328	214	0	57	48	31	16
t	259	2	31	1	583	0	2	3774	252	0	0	75	1	2
u	45	53	114	48	71	0	148	0	65	0	0	247	87	278
v	27	0	0	0	683	0	0	0	109	0	1	0	0	0
w	595	3	0	6	285	0	0	472	374	0	0	12	0	103
x	17	0	9	0	9	0	0	0	10	0	0	0	0	0
y	11	10	0	0	152	0	1	1	32	0	0	7	1	0
z	3	0	0	0	26	0	0	0	2	0	0	4	0	0
SPACE	1882	1033	864	515	423	1059	453	1388	237	93	152	717	876	478

1st/2nd	o	p	q	r	s	t	u	v	w	x	y	z	SPACE
a	1	100	0	802	683	785	87	233	57	0	319	12	50
b	240	0	0	88	15	7	256	1	1	0	13	0	36
c	298	0	0	71	7	154	34	0	0	0	9	0	47
d	137	0	0	83	95	3	52	5	2	0	51	0	2627
e	44	90	9	1314	630	316	8	172	106	0	189	2	4904
f	429	0	0	188	4	102	62	0	0	0	4	0	110
g	135	0	0	98	42	6	57	0	1	0	2	0	686
h	487	2	0	91	8	165	75	0	8	0	32	0	715
i	88	42	2	272	484	558	5	165	0	0	0	18	4
j	41	0	0	0	0	0	56	0	0	0	0	0	0
k	3	1	0	0	50	0	3	0	19	0	8	0	309
l	344	34	0	11	121	74	81	17	0	0	276	0	630
m	289	77	0	0	53	2	85	0	2	0	19	0	454
n	349	0	0	2	148	378	49	3	0	0	115	0	1152
o	336	138	0	812	195	415	1115	136	398	0	47	5	294
p	149	64	0	110	48	40	68	0	0	0	14	0	127
q	0	0	0	0	0	0	66	0	0	0	0	0	0
r	504	9	0	69	318	190	89	22	5	0	145	0	1483
s	213	107	8	0	168	754	175	0	32	0	34	0	2228
t	331	0	0	187	209	154	132	0	84	23	121	1	2343
u	3	49	1	402	299	492	0	0	0	0	7	3	255
v	33	0	0	0	0	0	1	1	0	0	11	0	0
w	264	0	0	35	21	4	2	2	0	0	0	0	326
x	1	22	0	0	0	23	8	8	0	0	0	0	21
y	339	16	0	0	81	2	1	1	0	0	0	9	1171
z	2	0	0	0	3	0	0	0	0	0	3	2	2
SPACE	721	588	42	494	1596	3912	134	116	1787	0	436	0	0

Figure A–3. Bi-gram frequency table B_2, from Konheim's introductory cryptography textbook, to which we have added the same space bi-gram extensions of Figure A–2. Because Konheim's total count of A to Z bi-grams (67,227) is nearly identical to that of Mayzner and Tresselt (67,320), we can conveniently add the space bi-grams with only minor renormalization.

1st/2nd	a	b	c	d	e	f	g	h	i	j	k	l	m	n
a	7	125	251	304	13	65	151	13	311	13	67	681	182	1216
b	114	7	2	1	394	0	0	0	74	7	0	152	6	0
c	319	0	52	1	453	0	0	0	202	0	86	98	4	3
d	158	3	4	33	572	20	1	20	273	0	0	19	27	8
e	492	27	323	890	187	93	106	93	118	0	27	340	253	1029
f	98	0	0	150	150	108	108	0	188	0	0	35	1	1
g	122	0	2	108	271	20	20	0	0	0	23	3	51	129
h	646	2	5	285	2053	174	0	0	426	0	0	6	6	14
i	236	51	476	130	271	0	145	95	10	0	31	352	184	1550
j	18	0	0	0	26	0	0	0	5	0	0	3	0	0
k	14	1	0	1	0	0	0	0	56	0	0	7	0	0
l	359	5	6	197	513	28	80	7	407	0	4	378	1	20
m	351	65	2	0	573	2	0	0	259	0	21	2	22	1
n	249	2	281	761	310	46	630	0	301	4	0	33	126	8
o	48	57	91	130	0	731	0	46	52	8	30	234	47	88
p	241	0	1	1	7	0	0	0	75	0	44	144	397	1232
q	0	0	0	0	0	0	0	0	0	0	0	0	13	1
r	470	15	79	129	1280	14	14	80	541	0	0	75	0	0
s	200	4	94	9	595	8	8	0	390	0	94	48	139	149
t	381	2	22	51	872	4	4	80	865	0	30	62	37	7
u	72	87	103	2	91	11	11	0	54	0	0	230	27	9
v	65	0	0	4	522	0	0	0	223	0	3	0	69	318
w	282	1	0	0	239	0	0	0	259	0	0	5	1	0
x	9	0	15	15	17	0	0	0	15	0	0	0	0	44
y	17	1	3	3	84	0	0	0	20	0	0	5	1	0
z	18	0	0	2	36	0	0	0	17	0	1	1	11	5
SPACE	1882	1033	864	515	423	1059	453	1388	237	93	152	717	876	478

1st/2nd	o	p	q	r	s	t	u	v	w	x	y	z	SPACE
a	5	144	0	764	648	1019	89	137	37	17	202	15	50
b	118	0	0	81	28	6	89	2	0	0	143	0	36
c	606	0	1	113	23	237	92	0	0	0	25	0	47
d	111	0	1	49	75	2	91	15	6	113	40	0	2627
e	30	143	25	1436	917	301	36	160	153	0	90	3	4904
f	326	0	0	142	3	58	54	0	0	6	5	0	110
g	0	0	150	29	28	58	0	0	0	0	0	0	686
h	287	0	0	56	10	85	31	0	4	14	15	0	715
i	554	62	5	212	741	704	7	155	0	0	1	49	4
j	45	0	0	1	0	0	48	0	0	0	0	0	0
k	7	0	0	3	39	1	1	0	0	0	4	0	309
l	208	11	0	9	104	68	72	15	3	0	219	0	630
m	240	139	0	5	47	1	65	1	0	1	37	0	454
n	239	2	3	5	340	743	56	31	8	7	71	2	1152
o	125	164	0	861	201	223	533	188	194	0	23	2	294
p	268	103	0	409	32	51	81	0	0	0	3	0	127
q	0	0	0	0	0	0	73	0	0	1	0	0	0
r	510	25	0	97	300	273	88	65	8	0	140	0	1483
s	234	128	3	9	277	823	192	0	13	0	27	0	2228
t	756	2	0	295	257	131	120	3	54	2	125	3	2343
u	4	81	0	306	256	263	6	3	0	0	3	1	255
v	46	0	0	0	2	0	1	1	0	0	5	0	0
w	159	0	0	13	45	2	0	0	0	5	3	0	326
x	1	47	0	0	0	23	4	0	0	0	0	0	21
y	64	9	0	9	44	5	1	0	0	0	2	1	1171
z	4	0	0	0	0	0	0	0	0	0	0	2	2
SPACE	721	588	42	494	1596	3912	134	116	1787	0	436	2	0

Figure A-4. Bi-gram frequency table B_3, our own table generated from a corpus 10 times the size of that used for B_1 and B_2, consisting of a mixture of formal and informal English (e-mail and classic novels).

1st/2nd	a	b	c	d	e	f	g	h	i	j	k	l	m	n
a	2	1218	1968	2368	11	382	894	31	2274	32	695	4401	1419	9584
b	636	58	11	7	3595	3	6	1	419	58	0	1219	61	3
c	2272	0	283	4	3022	10	2	2915	687	0	950	805	2	0
d	854	14	3	313	3561	39	178	5	1816	16	10	264	36	126
e	3647	211	1587	6084	2768	957	390	121	658	14	162	2843	1715	7279
f	1019	0	0	4	963	523	2	1	1380	0	0	280	1	5
g	571	22	6	6	1966	8	178	1716	653	0	0	336	35	176
h	7858	407	2	4	16596	9	0	0	5444	7	0	43	41	48
i	854	0	2903	2007	1314	814	1428	2	15	0	281	1930	1821	12853
j	43	1	0	8	154	0	0	3	0	2	0	0	0	0
k	50	30	0	3	1606	28	13	4	526	0	0	60	5	432
l	2484	360	26	1760	4571	395	151	1	2577	1	181	3711	53	9
m	2886	18	18	4	4331	12	1	23	1399	53	1	21	356	57
n	1183	452	1924	7392	3928	304	5759	67	1450	15	587	467	31	401
o	283	6	709	837	182	5093	422	296	714	0	581	1312	3360	8544
p	1373	0	4	16	2122	27	7	0	446	0	7	1144	32	0
q	0	0	0	0	0	0	0	0	22	0	0	0	0	0
r	2663	87	267	1373	9285	98	257	109	2860	12	505	295	716	721
s	1462	41	486	53	5033	111	19	1973	2564	14	273	279	342	94
t	2364	6	229	2	5364	25	1	18290	5137	0	3	810	62	48
u	571	570	884	249	611	89	858	1	547	0	15	2174	509	2087
v	534	4	570	0	5429	0	0	0	966	0	0	2	2	0
w	2620	0	4	17	2325	4	0	2772	2475	0	6	76	3	602
x	203	40	0	0	94	5	1	5	218	0	0	2	1	0
y	36	0	30	30	503	9	1	4	214	0	0	60	94	49
z	40	0	40	0	193	0	0	0	43	0	0	18	0	0
SPACE	15637	5825	5611	3963	3009	5348	2053	8139	8886	397	569	3364	6013	2993

1st/2nd	o	p	q	r	s	t	u	v	w	x	y	z	SPACE
a	24	1024	3	5042	5285	8404	559	1924	430	106	1715	52	4038
b	954	5	0	620	280	132	1082	59	1	0	652	0	85
c	3299	0	20	608	82	1745	742	0	0	0	132	11	278
d	1638	35	0	534	606	7	453	115	18	8	318	0	13169
e	276	774	225	10955	5514	2867	23	1360	526	960	616	29	25931
f	2731	2	0	1121	18	476	517	0	2	1	68	0	5246
g	815	0	0	712	216	48	340	0	0	0	56	1	3899
h	2824	0	0	443	50	1121	334	1150	6	0	194	0	3634
i	2950	305	76	1392	6474	6267	31	0	0	115	1	204	37
j	125	0	0	0	0	0	337	0	0	0	0	0	0
k	14	0	0	0	297	6	6	112	13	0	37	0	1283
l	1976	138	0	1	536	422	555	0	75	0	2487	1	3347
m	1626	1083	0	0	428	5	671	205	2	0	1305	0	2034
n	3137	41	32	2	1926	4524	294	704	29	15	566	6	9940
o	1927	927	2	6134	1359	316	7904	0	2519	54	118	13	6488
p	1761	651	0	1669	286	0	380	4	8	0	64	0	903
q	0	0	0	0	0	0	846	245	4	0	0	0	4
r	3621	282	0	2	1636	1397	629	3	133	0	1570	0	7160
s	2119	712	59	13	1982	5112	1275	2	200	0	227	0	13916
t	6280	15	0	1855	1239	1408	1043	5	589	0	759	5	14050
u	55	1328	0	3135	2305	2832	3	0	17	18	27	15	1621
v	159	0	0	1	1	0	8	0	0	0	30	0	23
w	1471	3	0	218	181	4	4	0	5	0	9	0	1303
x	4	232	1	1	0	221	3	0	4	2	11	0	150
y	2456	73	0	4	450	115	8	0	27	0	0	4	7491
z	5	0	0	0	0	0	12	0	0	2	4	25	26
SPACE	9289	3657	396	2787	9383	20491	1969	1028	9162	25	2548	16	8079

Human Factors Interventions for the Health Care of Older Adults

Edited by
Wendy A. Rogers • Arthur D. Fisk

HUMAN FACTORS INTERVENTIONS FOR THE HEALTH CARE OF OLDER ADULTS

Edited by
Wendy A. Rogers
Arthur D. Fisk
Georgia Institute of Technology

This book represents the culmination of discussions that took place at the Conference on Human Factors and Health Care Interventions for Older Adults. The relevance and contributions of the field of human factors to health care interventions is the unifying theme of this book. Initial chapters provide overviews of the field of human factors and the primary research methodologies of that field. Other chapters review the cognitive issues that must be considered in the context of the health care environment and the potential for exercises to improve such cognitive functions. The remaining chapters cover a range of cutting-edge topics including: care giving, telecommunication issues, design of medical devices, computer monitoring of patients, automated communication systems, computer interface issues in general, and the use of the World Wide Web as a source for health information. Written by experts in the field of aging or health care, this book will be of interest to researchers and practitioners in the fields of aging, human factors, and health care.

Lawrence Erlbaum Associates, Inc.
10 Industrial Ave., Mahwah, NJ 07430–2262
201–258–2200; 1–800–926–6579; fax 201–760–3735
orders@erlbaum.com; www.erlbaum.com

SUBSCRIPTION ORDER FORM

Please ☐ enter ☐ renew my subscription to:

HUMAN–COMPUTER INTERACTION

A JOURNAL OF THEORETICAL, EMPIRICAL, AND METHODOLOGICAL ISSUES OF USER SCIENCE AND SYSTEM DESIGN

Volume 17, 2002, Quarterly — ISSN 0737–0024/Online ISSN 1532–7051

SUBSCRIPTION PRICES PER VOLUME:

Individual:
☐ $55.00 US/Canada
☐ $85.00 All Other Countries

Institution:
☐ $395.00 US/Canada
☐ $425.00 All Other Countries

Online Only:
☐ $49.50 Individual
☐ $355.50 Institution

Subscriptions are entered on a calendar-year basis only and must be paid in advance in U.S. currency—check, credit card, or money order. Prices for subscriptions include postage and handling. Journal prices expire 12/31/02. **NOTE:** Institutions must pay institutional rates. Individual subscription orders are welcome if prepaid by credit card or personal check. **Online access is included with individual subscriptions.**

☐   **Check Enclosed** (U.S. Currency Only)   Total Amount Enclosed $_____

☐   **Charge My:** ☐ VISA ☐ MasterCard ☐ AMEX ☐ Discover

Card Number _____   Exp. Date____/____

Signature_____

(Credit card orders cannot be processed without your signature.)

PRINT CLEARLY for proper delivery. STREET ADDRESS/SUITE/ROOM # REQUIRED FOR DELIVERY.

Name_____

Address_____

City/State/ Zip+4 _____

Daytime Phone #_____E-mail address_____

Prices are subject to change without notice.

Please note: A $20.00 penalty will be charged against customers providing checks that must be returned for payment. This assessment will be made only in instances when problems in collecting funds are directly attributable to customer error.

For information about online subscriptions, visit our website at *www.erlbaum.com*

Mail orders to: Lawrence Erlbaum Associates, Inc., Journal Subscription Department
10 Industrial Avenue, Mahwah, NJ 07430; **(201) 258–2200; FAX (201) 760–3735; journals@erlbaum.com**

LIBRARY RECOMMENDATION FORM

Detach and forward to your librarian.

☐ I have reviewed the description of *Human–Computer Interaction* and would like to recommend it for acquisition.

HUMAN–COMPUTER INTERACTION

A JOURNAL OF THEORETICAL, EMPIRICAL, AND METHODOLOGICAL ISSUES OF USER SCIENCE AND SYSTEM DESIGN

Volume 17, 2002, Quarterly — ISSN 0737–0024/Online ISSN 1532–7051

Institutional Rate: ☐ **$395.00** (US & Canada)   ☐ **$425.00** (All Other Countries)

Name_____Title_____

Institution/Department_____

Address_____

E-Mail Address _____

Librarians, please send your orders directly to LEA or contact from your subscription agent.

Lawrence Erlbaum Associates, Inc., Journal Subscription Department
10 Industrial Avenue, Mahwah, NJ 07430; **(201) 258–2200; FAX (201) 760–3735; journals@erlbaum.com**

HANDBOOK OF VIRTUAL ENVIRONMENTS
Design, Implementation, and Applications
Edited by

Kay M. Stanney
University of Central Florida
A Volume in the Human Factors and Ergonomics Series
Series Editor: *Gavriel Salvendy*

This *Handbook* provides a comprehensive, state-of-the-art account of virtual environments (VE) and serves as an invaluable source of reference for practitioners, researchers, and students in this rapidly evolving discipline. It provides practitioners with a reference source to guide their development efforts and addresses technology concerns as well as the social and business implications with which those associated with the technology are likely to grapple. While each chapter has a strong theoretical foundation, practical implications are derived and illustrated via the many tables and figures presented throughout the book.

The *Handbook* presents a systematic and extensive coverage of the primary areas of research and development within VE technology. It brings together a comprehensive set of contributed articles that addresses the principles required to define system requirements and design, build, evaluate, implement, and manage the effective use of VE applications. The contributors provide critical insights and principles associated with their given area of expertise to provide extensive scope and detail on virtual environment technology.

0-8058-3270-X [cloth] / 2002 / 1272epp. / $295.00
Special Discount Price! $95.00
Applies if payment accompanies order or for course adoption orders of 5 or more copies.
No further discounts apply.
Prices are subject to change without notice.

Lawrence Erlbaum Associates, Inc.
10 Industrial Ave., Mahwah, NJ 07430–2262
201–258–2200; 1–800–926–6579; fax 201–760–3735
orders@erlbaum.com; www.erlbaum.com

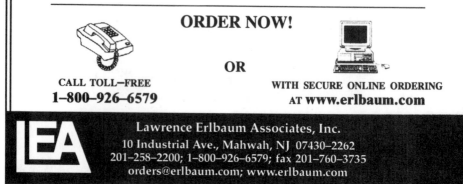

JOB OPPORTUNITIES

IOWA STATE UNIVERSITY
Human–Computer Interaction

Iowa State University invites applications for three tenured or tenure-track faculty positions in the field of Human Computer Interaction (HCI). As part of a newly announced Presidential initiative, the university seeks faculty members to enhance its growing academic and research program in HCI. The research component of the program will build on interdisciplinary work already under way at Iowa State University's Virtual Reality Applications Center.

Candidates must have a PhD or the terminal degree in their educational discipline. Successful applicants must have demonstrated research achievement in HCI and be able to establish an externally funded research group that develops national impact in HCI, and develop and teach courses that support the HCI graduate program. Senior level candidates must have an outstanding record of research achievement in the technical foundation of HCI as well as exceptional leadership capabilities. The tenured or tenure-track appointments will be made in one of the academic departments participating in the HCI graduate program.

Further information can be found at:
http://www.vrac.iastate.edu

Applicants should send an electronic letter of application and an attached resume including names and addresses of at least three references to:

Dr. J. Oliver, Chair of Search Commitee
Virtual Reality Applications Center
oliver@iastate.edu

---

*Review of applications will begin on **December 1, 2002.***
Iowa State University is an equal opportunity/affirmative action employer.

Handbook
of Human
Factors
Testing *and*
Evaluation

HANDBOOK OF HUMAN FACTORS TESTING AND EVALUATION

Second Edition

Edited by

Samuel G. Charlton
Waikato University

Thomas G. O'Brien
O'Brien and Associates

Like the first edition, the revision of this successful handbook responds to the growing need for specific tools and methods for testing and evaluating human-system interfaces. Indications are that the market for information on these tools and applications will continue to grow as we enter into the 21st century. One of the goals of offering a second edition is to expand and emphasize the application chapters, providing contemporary examples of human factors test and evaluation (HFTE) enterprises across a range of systems and environments. To be accurate, coverage of the standard tools and techniques used in HFTE have been updated as well.

New features of the *Handbook of Human Factors Testing and Evaluation*:
* New chapters have been added on human performance, usability, software usability, aviation and avionics, automotive and road safety, and applications of HFTE to industrial ergonomics.
* Updates regarding tools and methods are provided, including new testing methods that are used in Wiley's, *Handbook of Usability Testing* by Jeffrey Rubin.
* David Meister, "the most widely-cited author in the field of HFTE" is involved in updating this revision.
* More graphics and tables are incorporated.

The orientation of the current work has been toward breadth of coverage rather than in-depth treatment of a few issues or techniques. Experienced testers will find much that is familiar, as well as new tools, creative approaches, and a rekindled enthusiasm. Newcomers will discover the diversity of issues, methods, and creative approaches that make up the field. In addition, the book is written in such a way that individuals outside the profession should learn the intrinsic value and pleasure in ensuring safe, efficient, and effective operation, as well as increased user satisfaction through HFTE.

Contents: Preface. Part I: *Tools and Techniques*. Introduction. **T.G. O'Brien, D. Meister,** Human Factors Testing and Evaluation: An Historical Perspective. **S.G. Charlton, T.G. O'Brien,** The Role of Human Factors Testing and Evaluation in Systems Development. **S.G. Charlton,** Selecting Measures for Human Factors Tests. **T.G. O'Brien, T.B. Malone,** Test Documentation: Standards, Plans, and Reports. **D.H. Harris,** Human Performance Testing. **S.G. Charlton,** Measurement of Cognitive States in Test and Evaluation. **G.F. Wilson,** Psychophysiological Test Methods and Procedures. **B. Peacock,** Measurement in Manufacturing Ergonomics. **V.J. Gawron, T.W. Dennison, M.A. Biferno,** Mock-ups, Models, Simulations, and Embedded Testing. **S.G. Charlton,** Questionnaire Techniques for Test and Evaluation. **T.G. O'Brien,** Testing the Workplace Environment. **Part II: *Applications*.** Introduction to Applications Chapters. **A.R. Jacobsen, B.D. Kelly, D.L. Hilby, R.J. Mumaw,** Human Factors Testing and Evaluation in Commercial Airplane Flight Deck Development. **A.M. Rothblum, A.B. Carvalhais,** Maritime Applications of Human Factors Test and Evaluation. **R.J. Parker, T.A. Bentley, L. Ashby,** Human Factors Testing in the Forest Industry. **S.G. Charlton, B.D. Alley, P.H. Baas, J.E. Newman,** Human Factors Testing Issues in Road Safety. **J.M. O'Hara, W.F. Stubler, J.C. Higgins,** Human Factors Evaluation of Hybrid Human-System Interfaces in Nuclear Power Plants. **M. Gage, E.B-N. Sanders, C.T. William,** Generative Search in the Product Development Process. **S.D. Armstrong, W.C. Brewer, R.K. Steinberg,** Usability Testing. **L.S. Canham,** Operability Testing of Command, Control, Communications, Computers, and Intelligence (C4I) Systems. **A. Oudenhuijzen, P. Essens, T.B. Malone,** Ergonomics, Anthropometry, and Human Engineering. **J.M. Childs, H.H. Bell,** Training Systems Evaluation. **S.G. Charlton,** Epilogue: Testing Technology.
0-8058-3290-4 [cloth] / 2002 / 560pp. / $125.00
0-8058-3291-2 [paper] / 2002 / 560pp. / $59.95
Prices are subject to change without notice.

Lawrence Erlbaum Associates, Inc.
10 Industrial Ave., Mahwah, NJ 07430–2262
201–258–2200; 1–800–926–6579; fax 201–760–3735
orders@erlbaum.com; www.erlbaum.com